ALLE · ZEIT · WACH
1842

Friedo Giese

Beilstein's Index

Trivial Names in Systematic Nomenclature of Organic Chemistry

Springer-Verlag
Berlin Heidelberg New York Tokyo

Friedo Giese

Beilstein-Institut
Varrentrappstraße 40–42
D-6000 Frankfurt/Main 90

ISBN 3-540-16142-2 Springer-Verlag Berlin Heidelberg New York Tokyo
ISBN 0-387-16142-2 Springer-Verlag New York Heidelberg Berlin Tokyo

Library of Congress Cataloging-in-Publication Data
Giese, Friedo, Beilstein's Index. English and German. Includes indexes.
1. Chemistry, Organic – Nomenclature. I. Beilstein-Institut für Literatur der Organischen Chemie. II. Title.
QD291.G53 1986 547'.0014 85-27914
ISBN 0-387-16142-2 (U.S.)

Typesetting, Printing and Bookbinding: Konrad Triltsch, Graphischer Betrieb, Würzburg
2152/3145-543210

Foreword

Trivial names have always played an important role in the nomenclature of organic chemistry. Although originally meant to represent individual compounds some names have been adopted in systematic nomenclature as roots from which systematic names are derived by the addition of prefixes, affixes and suffixes. Many of the old trivial names have, however, been replaced by systematic names and they are no longer to be found in the current literature and are hardly known to the present-day chemist.

During the past few years a new generation of trivial names has been developed which allows compounds, particularly natural products with their large, and often complicated, skeletal ring systems, to be named more conveniently, than is possible with the rules of systematic organic nomenclature.

This book contains all the trivial names which are to be found as roots of names used in the chemical substance indexes of major reference works such as Beilstein's Handbook of Organic Chemistry and Chemical Abstracts, as well as all those trivial names which are allowed by the IUPAC Rules of Nomenclature. It is hoped that it will help all those who have problems in nomenclature to solve them rapidly and easily and so facilitate their use of the literature.

I should like to thank the directors of and my colleagues at the Beilstein Institute, Frankfurt/Main, for their generous support, criticism, and advice.

Frankfurt/Main, April 1986 Friedo Giese

Vorwort

Trivialnamen haben in der Nomenklatur der Organischen Chemie immer eine besondere Rolle gespielt. Ursprünglich nur als Bezeichnung für eine individuelle Verbindung gedacht, erhielten einige Namen in der sich entwickelnden systematischen Nomenklatur die Bedeutung als Stammnamen, von denen die systematischen Namen durch Zufügen von Präfixen und Suffixen abgeleitet wurden. Viele ältere Trivialnamen wurden aber durch systematische Namen ersetzt und verschwanden aus der Literatur, so daß sie den heutigen Chemikern kaum noch bekannt sind.

In den vergangenen Jahrzehnten ist eine neue Generation von Trivialnamen entwickelt worden, die insbesondere auf dem Gebiet der Naturstoffchemie mit ihren großen und teilweise auch komplizierten Ringgerüsten eine erheblich einfachere Benennung zulassen, als es mit den Regeln der systematischen Nomenklatur möglich wäre.

Dieses Büchlein enthält alle Trivialnamen, die als Stammnamen Eingang in die Sachregister der grossen Nachschlagewerke wie "Beilsteins Handbuch der Organischen Chemie" und "Chemical Abstracts" gefunden haben, sowie alle diejenigen Trivialnamen, die nach den Nomenklaturregeln der IUPAC noch als zulässig anerkannt sind. Möge es allen Chemikern einen schnelleren und leichteren Zugang zur Nomenklatur und damit zur Literatur ermöglichen.

Für vielfältige Unterstützung, Kritik und Beratung danke ich dem Vorstand und den Kollegen des Beilstein-Instituts, Frankfurt/M.

Frankfurt/M., im April 1986 Friedo Giese

Table of Contents

Trivialnames from A–Z 1

Root modifying prefixes 193

Stammverändernde Präfixe 201

List of IUPAC-rules 209

RingCode-Index/Register 211

Formula-Index 227

Reg.No.-Index 241

Introduction

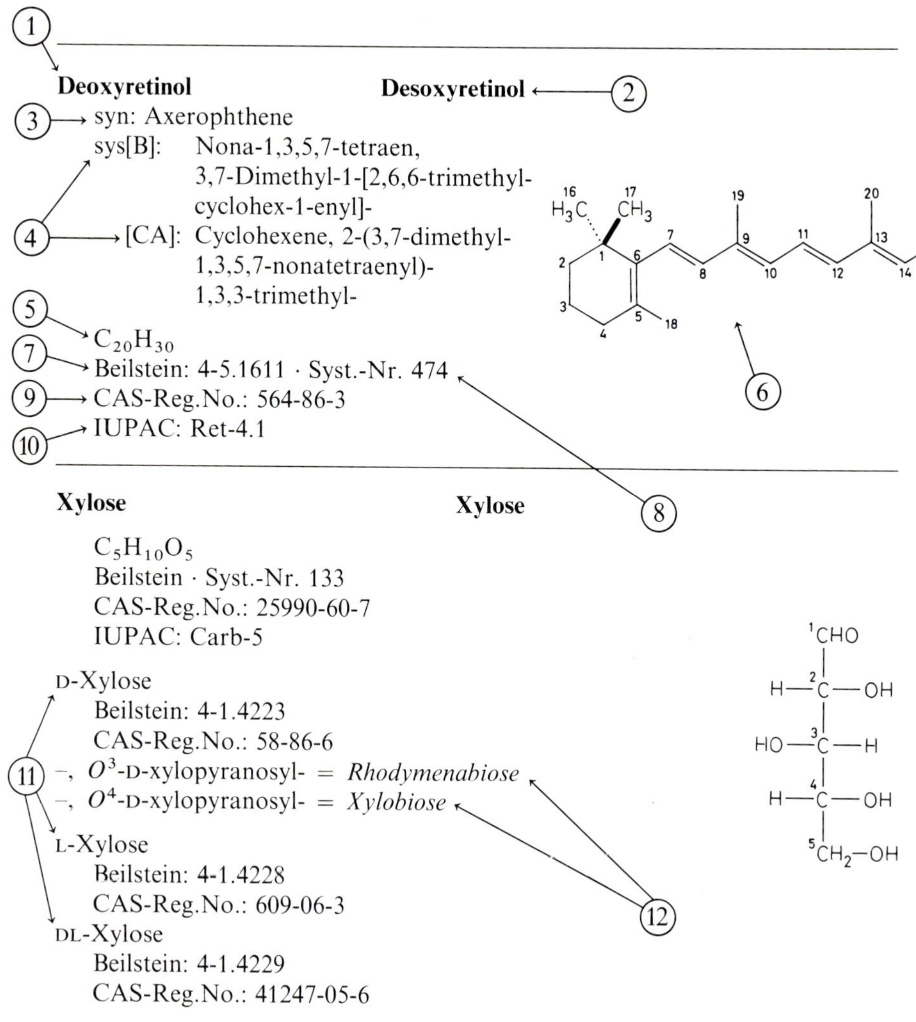

Deoxyretinol **Desoxyretinol**

syn: Axerophthene

sys[B]: Nona-1,3,5,7-tetraen, 3,7-Dimethyl-1-[2,6,6-trimethyl-cyclohex-1-enyl]-

[CA]: Cyclohexene, 2-(3,7-dimethyl-1,3,5,7-nonatetraenyl)-1,3,3-trimethyl-

$C_{20}H_{30}$

Beilstein: 4-5.1611 · Syst.-Nr. 474

CAS-Reg.No.: 564-86-3

IUPAC: Ret-4.1

Xylose **Xylose**

$C_5H_{10}O_5$

Beilstein · Syst.-Nr. 133

CAS-Reg.No.: 25990-60-7

IUPAC: Carb-5

D-Xylose

Beilstein: 4-1.4223

CAS-Reg.No.: 58-86-6

–, O^3-D-xylopyranosyl- = *Rhodymenabiose*

–, O^4-D-xylopyranosyl- = *Xylobiose*

L-Xylose

Beilstein: 4-1.4228

CAS-Reg.No.: 609-06-3

DL-Xylose

Beilstein: 4-1.4229

CAS-Reg.No.: 41247-05-6

Each entry contains, when available, the following information:

1) **Trivial name (English)**

2) **Trivial name (German).** The entries are arranged alphabetically according to their English names. German trivial names which differ considerably in their spelling from English are also arranged alphabetically, together with a cross reference. indication as to their position is provided.

3) **Synonyms.** Other trivial names which have occasionally been used in the literature are also given.

4) **Systematic names.** A systematic name is only given when the trivial name is not used by Beilstein [B] or Chemical abstracts [CA]. The systematic name is given as it appears in the appropriate index.

5) **Molecular formula**

6) **Structural formula.** As a rule the structural formula repeats the numbering used in the trivial name. An indication is given if and when numbering has been changed at any time or if IUPAC, Beilstein or Chemical Abstracts use different numbering.

7) **Beilstein.** Beilstein references are constructed as follows: the letter or number in front of the hyphen indicates either the Basic series (H-) or the appropriate Supplementary Series (1–5). The combined 3rd and 4th Supplementary Series is shown only as "4-". The single or double figures which follow indicate the volume number. In the 5th supplementary series the subvolume is also indicated separated from the volume number by a stroke since the subvolumes are paginated independently. The page number follows the full-stop. The reference "5-17/1.251" therefore means: 5th Supplementary Series, Volume 17, Subvolume 1, page 251. An asterik after the page number indicates that here there is no entry but only the trivial name.

8) **Syst.-Nr.** The Beilstein system number serves as an aid to searching in the Beilstein Handbook. It is printed on the book spines.

9) **CAS-Reg.No.** The Chemical Abstracts Service Registry Number is a serially assigned number for each compound described in CA since 1965. Knowing the Reg.No. can speed up on-line searches considerably.

10) **IUPAC.** The letter(s) in front of the hyphen designates the appropriate set of IUPAC rules (see List of IUPAC Rules, page 209) and the number after the hyphen the appropriate rule. A number in parentheses is the formula number in the corresponding IUPAC rule.

11) **Isomers and Stereoisomers.** The Beilstein reference and CAS Reg.No. are given for individual isomers or stereoisomers.

12) **Notes.** When both a (semi-)systematic name and a trivial name are shown together the trivial name is preferred when it is printed normally; the systematic name takes precedence when the trivial name appears in italics.

Einleitung

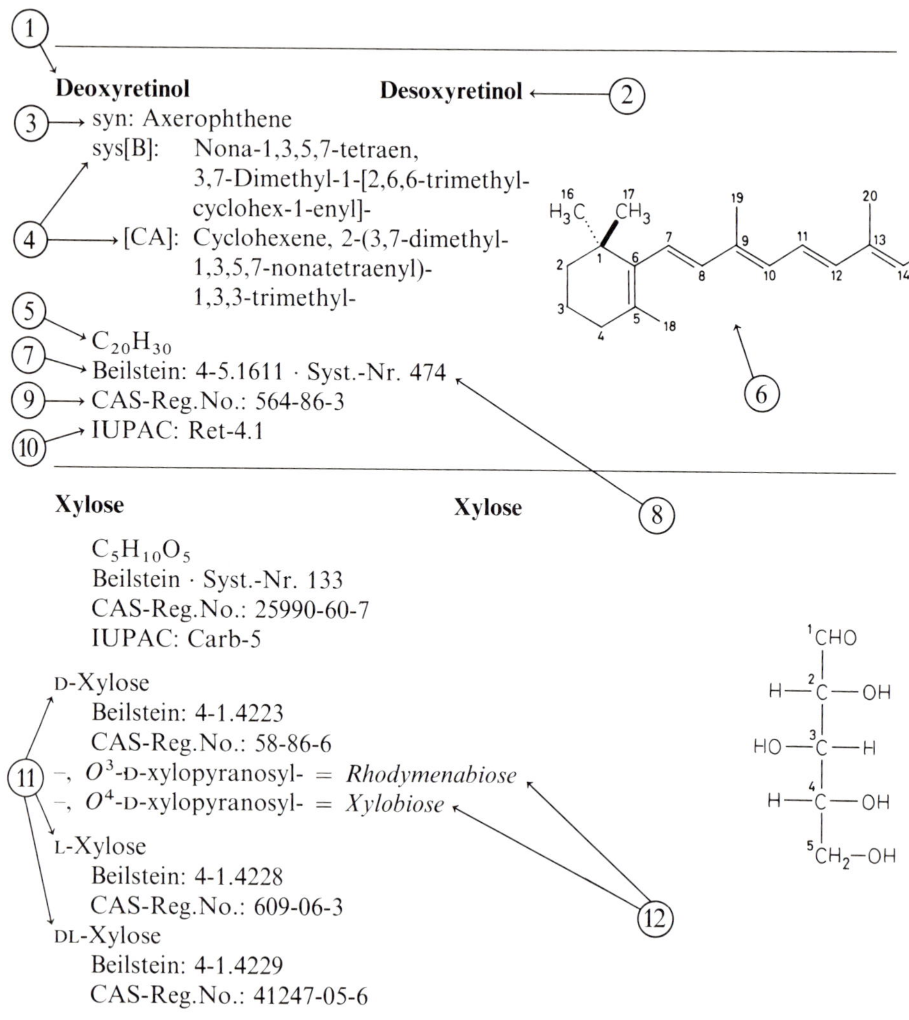

(1)

Deoxyretinol **Desoxyretinol** ← (2)

(3) → syn: Axerophthene

sys[B]: Nona-1,3,5,7-tetraen, 3,7-Dimethyl-1-[2,6,6-trimethyl-cyclohex-1-enyl]-

(4) → [CA]: Cyclohexene, 2-(3,7-dimethyl-1,3,5,7-nonatetraenyl)-1,3,3-trimethyl-

(5) → $C_{20}H_{30}$

(7) → Beilstein: 4-5.1611 · Syst.-Nr. 474 ← (8)

(9) → CAS-Reg.No.: 564-86-3

(10) → IUPAC: Ret-4.1

(6)

Xylose **Xylose**

$C_5H_{10}O_5$
Beilstein · Syst.-Nr. 133
CAS-Reg.No.: 25990-60-7
IUPAC: Carb-5

(11)

D-Xylose
Beilstein: 4-1.4223
CAS-Reg.No.: 58-86-6
–, O^3-D-xylopyranosyl- = *Rhodymenabiose*
–, O^4-D-xylopyranosyl- = *Xylobiose* ← (12)

L-Xylose
Beilstein: 4-1.4228
CAS-Reg.No.: 609-06-3

DL-Xylose
Beilstein: 4-1.4229
CAS-Reg.No.: 41247-05-6

Jeder Artikel enthält, soweit vorhanden, die folgenden Angaben:

1) **Trivialname (in Englisch)**

2) **Trivialname (in Deutsch)**. Die Artikel sind nach dem englischen Trivialnamen alphabetisch geordnet. Bei deutschen Trivialnamen, die erheblich in ihrer Schreibweise abweichen, ist durch einen alphabetisch eingeordneten Hinweis für ihre Auffindbarkeit gesorgt.

3) **Synonyma**. An dieser Stelle werden weitere Trivialnamen erwähnt, die gelegentlich für diese Verbindung in der Literatur verwendet worden sind.

4) **Systematische Namen**. Ein systematischer Name wird nur dann angegeben, wenn von BEILSTEIN [B] oder CHEMICAL ABSTRACTS [CA] der Trivialname nicht verwendet wird. Der Eintrag des systematischen Namens erfolgt in der Form und Schreibweise, wie er in den entsprechenden Sachregistern benutzt wird.

5) **Summenformel**

6) **Strukturformel**. Die Strukturformel gibt in der Regel die Bezifferung des Trivialnamens wieder. Hinweise sind gegeben, wenn sich die Bezifferung im Laufe der Zeit geändert hat, oder wenn von IUPAC, Chemical Abstracts oder Beilstein unterschiedliche Bezifferungen benutzt werden.

7) **Beilstein**. Das Beilstein-Zitat ist wie folgt zu lesen: die Zahl vor dem Bindestrich gibt das Hauptwerk (H-) oder das entsprechende Ergänzungswerk (1- bis 5-) an. Das vereinigte 3./4. Ergänzungswerk wird nur mit "4-" angegeben. Die folgende ein- oder zweistellige Zahl gibt den jeweiligen Band an. Beim 5. Ergänzungswerk ist auch eine Angabe des Teilbandes (von der Bandzahl durch einen Schrägstrich getrennt) erforderlich, da die einzelnen Teilbände getrennt paginiert werden. Nach dem Punkt folgt die Seitenzahl. Das Zitat "5-17/1.251" bedeutet also: 5. Ergänzungswerk, Band 17, Teilband 1, Seite 251.
Ein Stern hinter der Seitenzahl weist darauf hin, daß an dieser Stelle kein Artikel, sondern lediglich ein Hinweis auf den Trivialnamen zu finden ist.

8) **Syst.-Nr.** Die Beilstein-System-Nr. dient als Suchhilfe im Beilstein-Handbuch. Sie ist auf dem Buchrücken jedes Beilstein-Bandes wiedergeben.

9) **CAS-Reg.No.** Die Registry Number des Chemical Abstracts Service ist eine willkürlich vergebene Nummer für jede seit 1965 im CA beschriebene Verbindung. Bei Kenntnis der Reg.No. können insbesondere on-line-Recherchen schneller gestaltet werden.

10) **IUPAC**. Die Buchstabenfolge vor dem Bindestrich weist auf das entsprechende IUPAC-Regelwerk hin (s. List of IUPAC-Rules, Seite 209), die folgende Zahl auf die individuell zutreffende Regel. Eine in Klammern nachgestellte Zahl gibt die Formelnummer in der entsprechenden IUPAC-Regel wieder.

11) **Isomere und Stereoisomere**. Hier werden Beilstein-Zitat und CAS-Reg.No. für individuelle Isomere oder Stereoisomere gegeben.

12) **Hinweise**. Wird von einem (halb-)systematischen Namen auf einen Trivialnamen verwiesen, so ist der Trivialname zu bevorzugen, wenn er im normalen Druckbild erscheint; der systematische Name hat Vorrang, wenn der Trivialname kursiv gedruckt ist.

Trivialnames from A–Z

Abequose **Abequose**

= D-*xylo*-3,6-Dideoxy-hexose

Abietane **Abietan**

sys[CA]: Phenanthrene, tetradecahydro-1,1,4a-trimethyl-7-(1-methylethyl)-, (4aα,4bβ,7β,8aα,10aβ)-

$C_{20}H_{36}$
Beilstein: 3-5.422 · Syst.-Nr. 461
IUPAC: F-2

(4aR)-Abietane
CAS-Reg.No.: 19407-12-6
(4aS)-Abietane
CAS-Reg.No.: 18375-16-1

Aceanthrene **Aceanthren**

sys[CA]: Aceanthrylene, 1,2-dihydro-

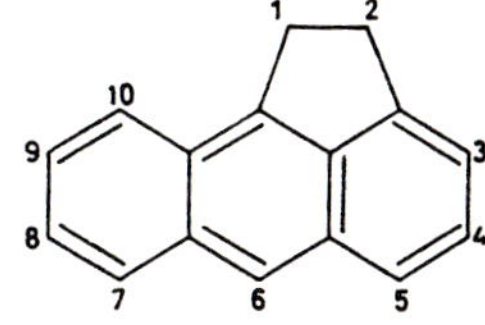

$C_{16}H_{12}$
Beilstein: 4-5.2416 · Syst.-Nr. 486
CAS-Reg.No.: 641-48-5
IUPAC: A-23.1

Aceanthrylene **Aceanthrylen**

$C_{16}H_{10}$
Beilstein · Syst.-Nr. 487
CAS-Reg.No.: 202-03-9
IUPAC: A-21.2(16)

–, 1,2-dihydro- – Aceanthrene

Acenaphthene **Acenaphthen**

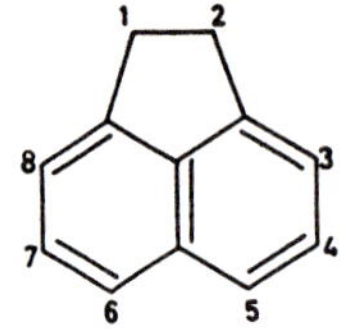

sys[CA]: Acenaphthylene, 1,2-dihydro-

$C_{12}H_{10}$
Beilstein: 4-5.1834 · Syst.-Nr. 479
CAS-Reg.No.: 83-32-9
IUPAC: A-23.1

Acenaphthylene **Acenaphthylen**

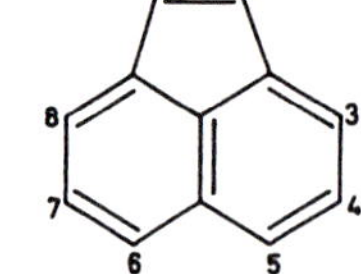

$C_{12}H_8$
Beilstein: 4-5.2138 · Syst.-Nr. 480
CAS-Reg.No.: 208-96-8
IUPAC: A-21.2(9)

–, 1,2-dihydro- = Acenaphthene

Acephenanthrene **Acephenanthren**

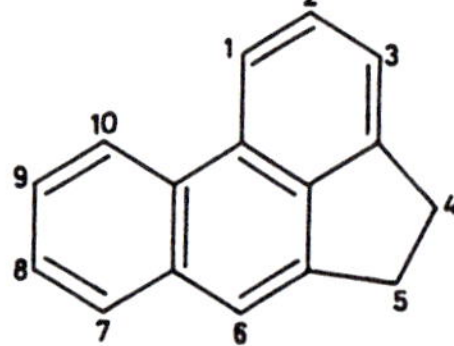

sys[CA]: Acephenanthrylene, 4,5-dihydro-

$C_{16}H_{12}$
Beilstein: 3-5.2235 · Syst.-Nr. 486
CAS-Reg.No.: 6232-48-0
IUPAC: A-23.1

Acephenanthrylene **Acephenanthrylen**

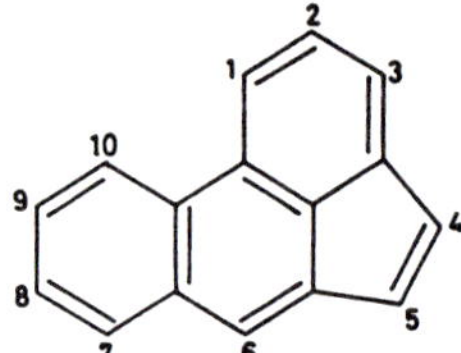

$C_{16}H_{10}$
Beilstein · Syst.-Nr. 487
CAS-Reg.No.: 201-06-9
IUPAC: A-21.2(15)

–, 4,5-dihydro- = Acephenanthrene

Acetaldehyde **Acetaldehyd**

C_2H_4O CH_3-CHO
Beilstein: 4-1.3094 · Syst.-Nr. 77
CAS-Reg.No.: 75-07-0
IUPAC: C-305.1

Acetic acid **Essigsäure**

$C_2H_4O_2$ $CH_3-CO-OH$
Beilstein: 4-2.94 · Syst.-Nr. 158
CAS-Reg.No.: 64-19-7
IUPAC: C-404.1

Acetoacetic acid **Acetessigsäure**
sys[B]: Buttersäure, 3-Oxo-
[CA]: Butanoic acid, 3-oxo-

$C_4H_6O_3$ $CH_3-CO-CH_2-CO-OH$
Beilstein: 4-3.1527 · Syst.-Nr. 280
CAS-Reg.No.: 541-50-4
IUPAC: C-416.3

Acetone **Aceton**
sys[CA]: 2-Propanone

C_3H_6O $CH_3-CO-CH_3$
Beilstein: 4-1.3180 · Syst.-Nr. 83
CAS-Reg.No.: 67-64-1
IUPAC: C-312.1

Acetylene **Acetylen**
sys[CA]: Ethyne

C_2H_2 $HC\equiv CH$
Beilstein: 4-1.939 · Syst.-Nr. 12
CAS-Reg.No.: 74-86-2
IUPAC: A-3.2

Acofriose **Acofriose**

= Mannose, O^3-methyl-6-deoxy-

Aconitane **Aconitan**

$C_{18}H_{27}N$
Beilstein: 4-21.2672* · Syst.Nr. 3073
CAS-Reg.No.: 34709-90-5
IUPAC: F-2

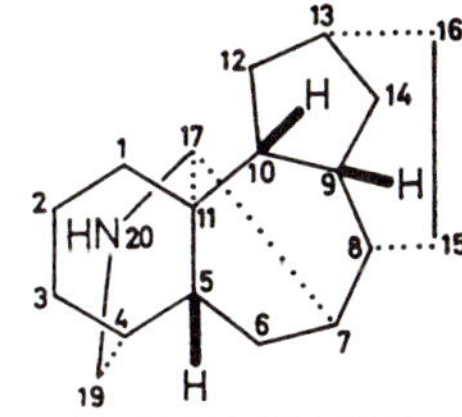

Acovenose **Acovenose**

= Talose, O^3-methyl-6-deoxy-

Acridarsine **Acridarsin**

$C_{13}H_9As$
Beilstein · Syst.-Nr. 4720
CAS-Reg.No.: 260-23-1

Acridine **Acridin**

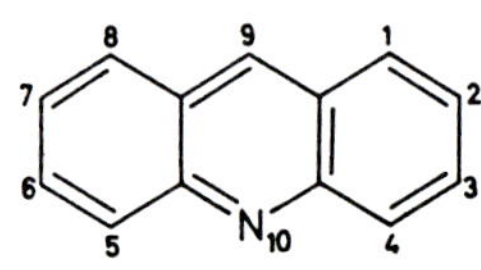

$C_{13}H_9N$
Beilstein: 4-20.3987 · Syst.-Nr. 3088
CAS-Reg.No.: 260-94-6
IUPAC: B-2.11(38)

Acrindoline **Acrindolin**
sys[B]: Indolo[3,2-*c*]acridin

$C_{19}H_{12}N_2$
Beilstein · Syst.Nr. 3492
CAS-Reg.No.: 42127-84-4

Acrylic acid **Acrylsäure**
sys[CA]: 2-Propenoic acid

$C_3H_4O_2$
Beilstein: 4-2.1455 · Syst.-Nr. 163
CAS-Reg.No.: 79-10-7
IUPAC: C-404.1

$CH_2{=}CH-CO-OH$

Acutumine **Acutumin**

$C_{19}H_{24}ClNO_6$
Beilstein: 4-21.6756 · Syst.-Nr. 3241
CAS-Reg.No.: 17088-50-5

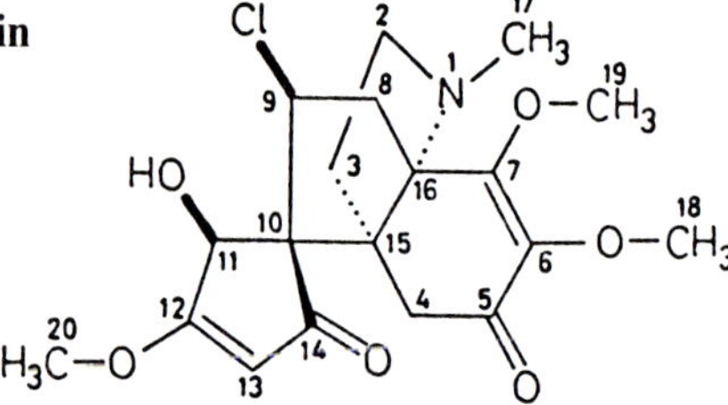

Adamantane **Adamantan**
sys[CA]: Tricyclo[3.3.1.1^{3,7}]decane

$C_{10}H_{16}$
Beilstein: 4-5.469 · Syst.-Nr. 459
CAS-Reg.No.: 281-23-2

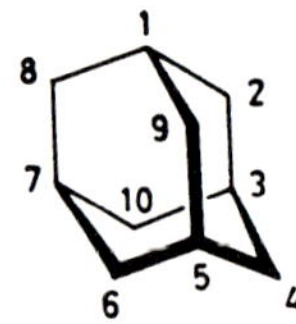

Adenine **Adenin**

sys[B]: 7(9)*H*-Purin-6-ylamin
[CA]: 1*H*-Purin-6-amine

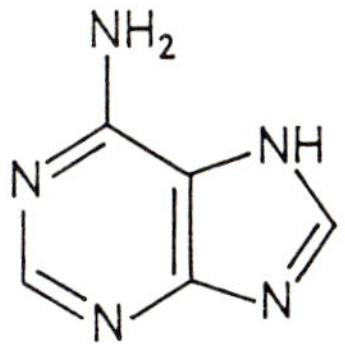

$C_5H_5N_5$
Beilstein: 4-26.3561 · Syst.-Nr. 4176
CAS-Reg.No.: 73-24-5
IUPAC: C-812.2

Adenosine **Adenosin**

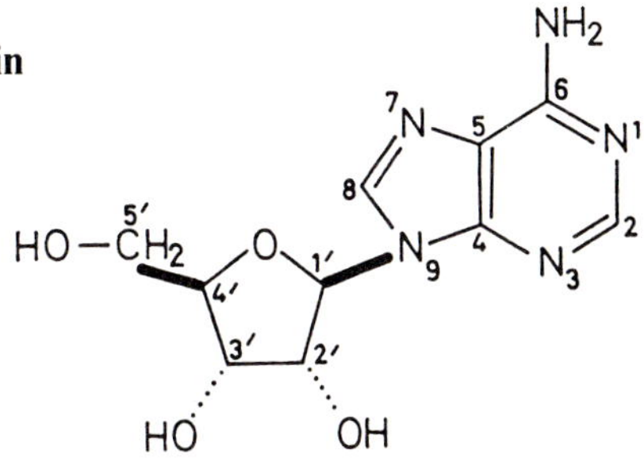

$C_{10}H_{13}N_5O_4$
Beilstein: 4-26.3598 · Syst.-Nr. 4176
CAS-Reg.No.: 58-61-7

Adermine **Adermin**

= Pyridoxine

Adipic acid **Adipinsäure**

sys[CA]: Hexanedioic acid

$C_6H_{10}O_4$
Beilstein: 4-2.1956 · Syst.-Nr. 175
CAS-Reg.No.: 124-04-9
IUPAC: C-404.1

$HO-CO-[CH_2]_4-CO-OH$

Äpfelsäure = Malic acid

Äthan = Ethane

Äthylenglykol = Ethylene glycol

Ätioporphyrin-I = Etioporphyrin I

Ajmalan **Ajmalan**

$C_{20}H_{26}N_2$
Beilstein: 4-23.2513* · Syst.-Nr. 3486
CAS-Reg.No.: 34612-39-0
IUPAC: F-2

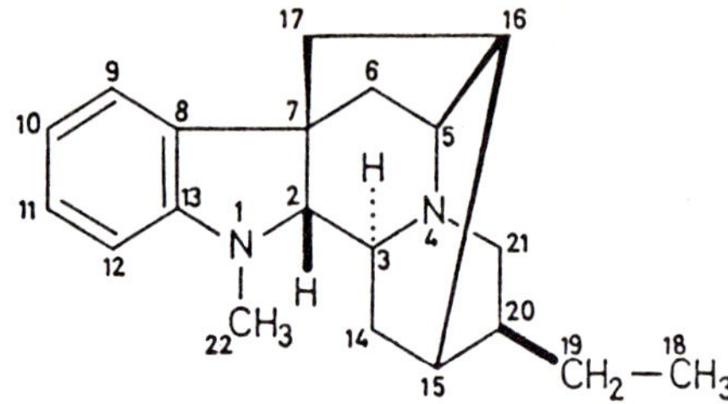

Akuammilan **Akuammilan**

$C_{19}H_{22}N_2$
Beilstein: 4-25.1258* · Syst.-Nr. 3487
CAS-Reg.No.: 5876-24-4
IUPAC: F-2

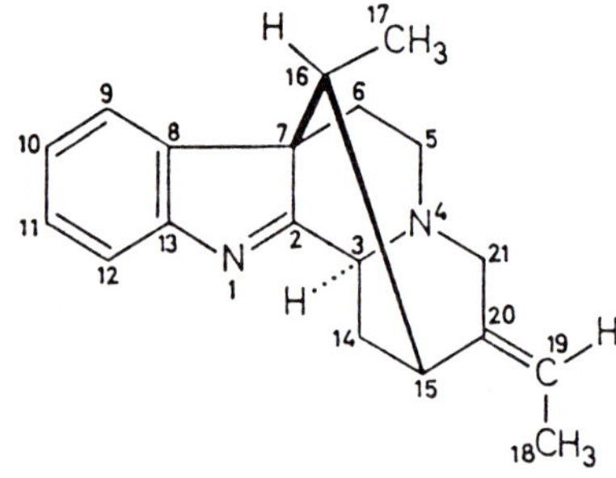

Alanine **Alanin**

$C_3H_7NO_2$
Beilstein · Syst.-Nr. 365 A
CAS-Reg.No.: 6898-94-8
IUPAC: 3AA-1

$H_3C-CH(NH_2)-CO-OH$

D-Alanine
Beilstein: 4-4.2480
CAS-Reg.No.: 338-69-2

L-Alanine
Beilstein: 4-4.2480
CAS-Reg.No.: 56-41-7

DL-Alanine
Beilstein: 4-4.2481
CAS-Reg.No.: 302-72-7

β-Alanine **β-Alanin**

sys[B]: Propionsäure, 3-Amino-

$C_3H_7NO_2$
Beilstein: 4-4.2526 · Syst.-Nr. 365 B
CAS-Reg.No.: 107-95-9
IUPAC: C-421.1; 3AA-Appendix

$H_2N-CH_2-CH_2-CO-OH$

Aldosterone **Aldosteron**

= Pregn-4-en-18-al, 11β,21-dihydroxy-3,20-dioxo-

Allene **Allen**

sys[CA]: 1,2-Propadiene

C_3H_4
Beilstein: 4-1.966 · Syst.-Nr. 12
CAS-Reg.No.: 463-49-0
IUPAC: A-3.1

$H_2C{=}C{=}CH_2$

Allocholane **Allocholan**

= Cholane, (5α)-

Allofuranose **Allofuranose**

$C_6H_{12}O_6$
Beilstein · Syst.-Nr. 144
IUPAC: Carb-5/18

α-D-Allofuranose
Beilstein: 4-1.4299
CAS-Reg.No.: 36468-79-8

β-D-Allofuranose
Beilstein: 4-1.4299
CAS-Reg.No.: 36468-80-1

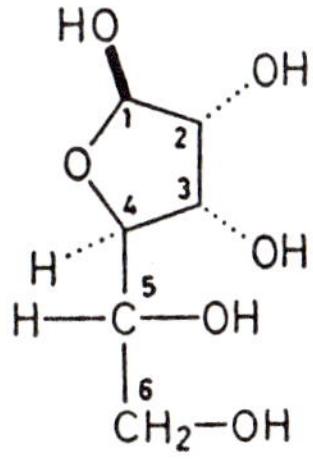

β-D-

Alloisoleucine **Alloisoleucin**

$C_6H_{13}NO_2$
CAS-Reg.No.: 26524-12-9
Beilstein · Syst.-Nr. 368
IUPAC: 3AA-4.4

D-Alloisoleucine
Beilstein: 4-4.2777
CAS-Reg.No.: 1509-35-9

L-Alloisoleucine
Beilstein: 4-4.2777
CAS-Reg.No.: 1509-34-8

DL-Alloisoleucine
Beilstein: 4-4.2777
CAS-Reg.No.: 3107-04-8

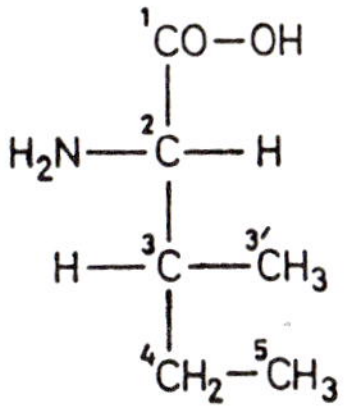

L-

Allomatridine **Allomatridin**

= Matridine, (6β)-

Allophanic acid **Allophansäure**

sys[CA]: Carbamic acid, (aminocarbonyl)-

$C_2H_4N_2O_3$
Beilstein: 3-3.134 · Syst.-Nr. 205 A
CAS-Reg.No.: 625-78-5
IUPAC: C-971.2

$H_2N-CO-NH-CO-OH$

Allopregnane **Allopregnan**

= Pregnane, (5α)-

Allopyranose **Allopyranose**

$C_6H_{12}O_6$
Beilstein · Syst.-Nr. 144
IUPAC: Carb-5/18

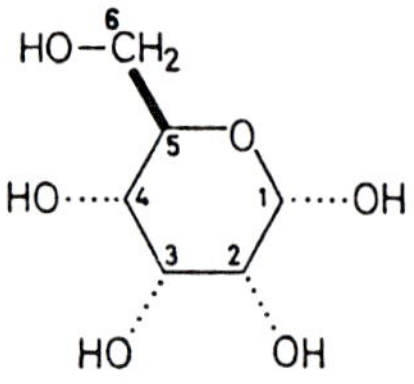

α-D-Allopyranose
Beilstein: 4-1.4299
CAS-Reg.No.: 7282-79-3

β-D-Allopyranose
Beilstein: 4-1.4299
CAS-Reg.No.: 7283-09-2

α-L-Allopyranose
CAS-Reg.No.: 39392-61-5

β-L-Allopyranose
Beilstein: 4-1.4300
CAS-Reg.No.: 39392-62-6

α-D-

Allose **Allose**

$C_6H_{12}O_6$
Beilstein · Syst.-Nr. 144
CAS-Reg.No.: 6038-51-3
IUPAC: Carb-5

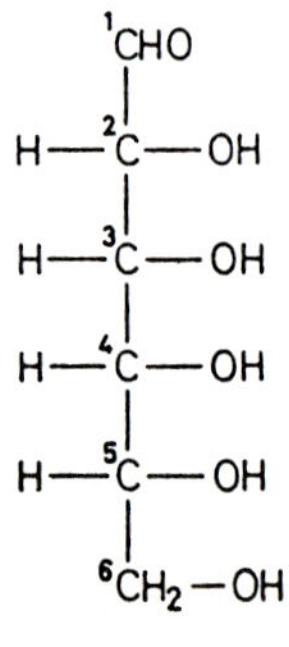

D-Allose
Beilstein: 4-1.4299
CAS-Reg.No.: 2595-97-3

L-Allose
Beilstein: 4-1.4300
CAS-Reg.No.: 7635-11-2

DL-Allose
Beilstein: 4-1.4300

D-

Allothreonine **Allothreonin**

$C_4H_9NO_3$
Beilstein · Syst.-Nr. 376
CAS-Reg.No.: 2676-21-3
IUPAC: 3AA-4.4

D-Allothreonine
Beilstein: 4-4.3170
CAS-Reg.No.: 24830-94-2
L-Allothreonine
Beilstein: 4-4.3170
CAS-Reg.No.: 28954-12-3
DL-Allothreonine
Beilstein: 4-4.3171
CAS-Reg.No.: 144-98-9

L-

Alloxan **Alloxan**
sys[CA]: 2,4,5,6(1*H*,3*H*)-Pyrimidine-tetrone

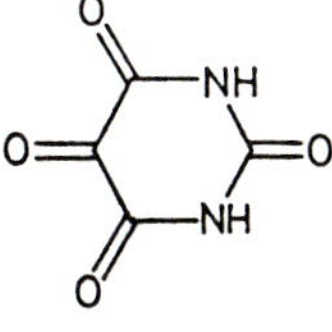

$C_4H_2N_2O_4$
Beilstein: 4-24.2137 · Syst.-Nr. 3627
CAS-Reg.No.: 50-71-5

Allysine **Allysin**

= Norleucine, 6-oxo-

Alnusane **Alnusan**

= Oleanane, *D:B*-friedo-

Alstophyllan **Alstophyllan**

$C_{21}H_{26}N_2O$
Beilstein: 4-25.169* · Syst.-Nr. 4495
CAS-Reg.No.: 38989-96-7
IUPAC: F-2

Altrofuranose **Altrofuranose**

$C_6H_{12}O_6$
Beilstein · Syst.-Nr. 144
IUPAC: Carb-5/18

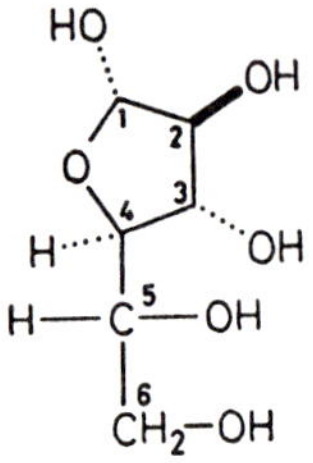

α-D-Altrofuranose
Beilstein: 4-1.4301
CAS-Reg.No.: 41846-93-9
β-D-Altrofuranose
Beilstein: 4-1.4301
CAS-Reg.No.: 40461-79-8
α-L-Altrofuranose
CAS-Reg.No.: 41846-91-7
β-L-Altrofuranose
CAS-Reg.No.: 36468-81-2

α-D-

Altropyranose **Altropyranose**

$C_6H_{12}O_6$
Beilstein · Syst.-Nr. 144
IUPAC: Carb-5/18

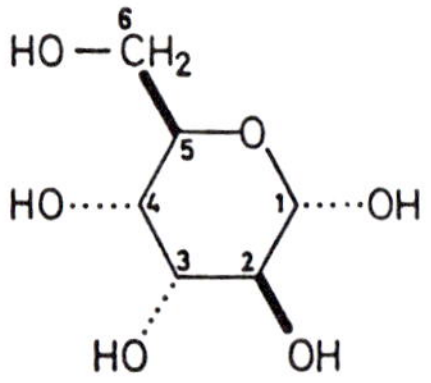

α-D-Altropyranose
Beilstein: 4-1.4301
CAS-Reg.No.: 7282-80-6
β-D-Altropyranose
Beilstein: 4-1.4301
CAS-Reg.No.: 7283-10-5
α-L-Altropyranose
CAS-Reg.No.: 12773-29-4
β-L-Altropyranose
Beilstein: 4-1.4301
CAS-Reg.No.: 12773-31-8

α-D-

Altrose **Altrose**

$C_6H_{12}O_6$
Beilstein · Syst.-Nr. 144
CAS-Reg.No.: 5987-68-8
IUPAC: Carb-5

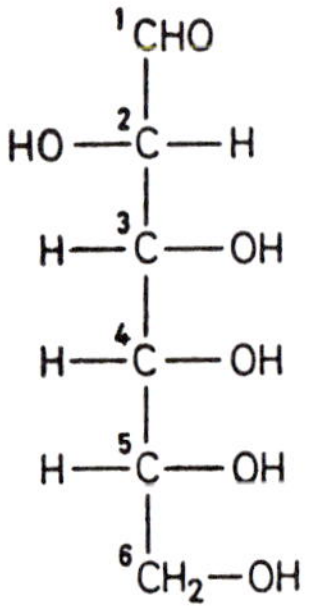

D-Altrose
Beilstein: 4-1.4300
CAS-Reg.No.: 1990-29-0
L-Altrose
Beilstein: 4-1.4301
CAS-Reg.No.: 1949-88-8

D-

Ambrane **Ambran**

sys[B]: Gammaceran, 8,14;13,18-Diseco-

[CA]: Naphthalene, decahydro-1,1,4a,6-tetramethyl-5-[4-methyl-6-(2,2,6-trimethylcyclohexyl)hexyl]-

$C_{30}H_{56}$
Beilstein: 3-5.1440* · Syst.-Nr. 461
CAS-Reg.No.: 468-83-7
IUPAC: F-(50)

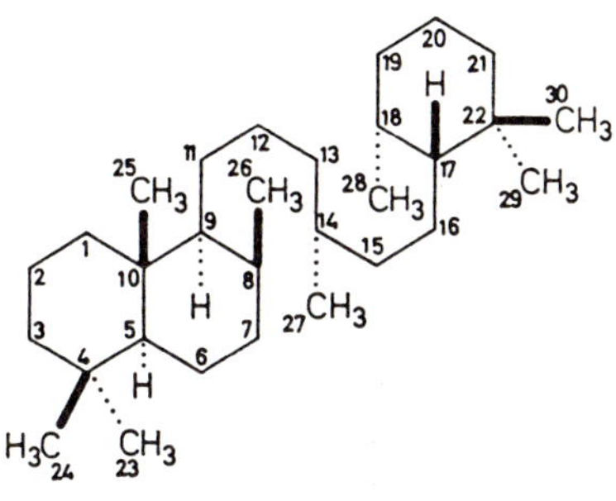

Ambrosane **Ambrosan**

sys[CA]: Azulene, decahydro-3a,8-dimethyl-5-(1-methylethyl)-, [3a*S*-(3aα,5α,8α,8aβ)]-

$C_{15}H_{28}$
Beilstein · Syst.-Nr. 453
CAS-Reg.No.: 24749-18-6
IUPAC: F-(6)

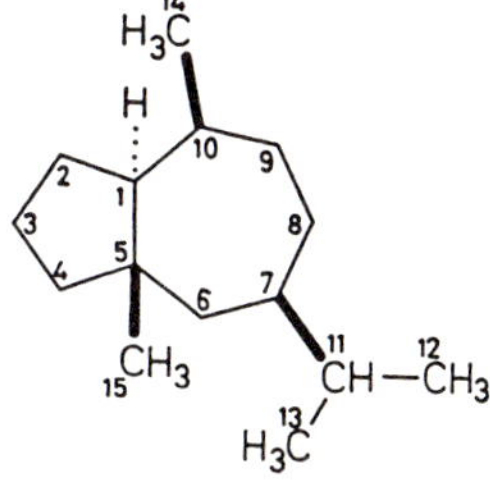

Ameisensäure = Formic acid

Amicetose **Amicetose**

= Hexanal, *erythro*-4,5-dihydroxy-

Anabasine **Anabasin**

sys[CA]: Pyridine, 3-(2-piperidinyl)-

$C_{10}H_{14}N_2$
Beilstein · Syst.-Nr. 3470
CAS-Reg.No.: 40774-73-0

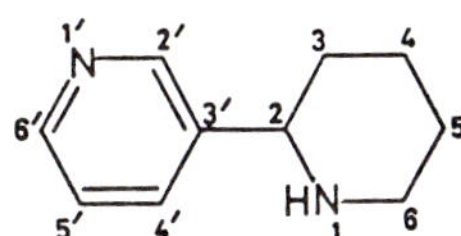

(+)-Anabasine
Beilstein: 4-23.1023

(−)-Anabasine
Beilstein: 4-23.1023
CAS-Reg.No.: 494-52-0

(±)-Anabasine
Beilstein: 4-23.1027
CAS-Reg.No.: 13078-04-1

Androstane **Androstan**

$C_{19}H_{32}$
Beilstein · Syst.-Nr. 472
CAS-Reg.No.: 24887-75-0
IUPAC: 2S-2.3(20)

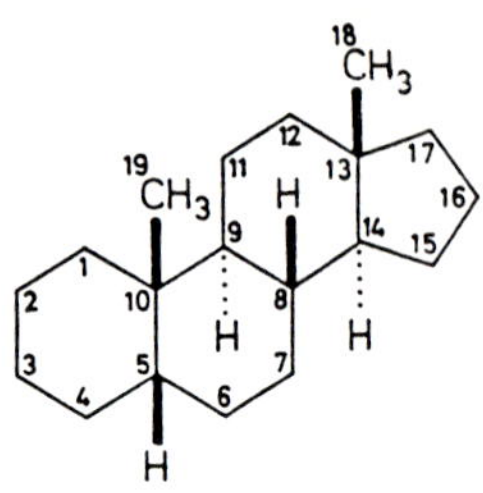

(5β)-

(5α)-Androstane
Beilstein: 4-5.1211
CAS-Reg.No.: 438-22-2
(5β)-Androstane [syn: Testane]
Beilstein: 4-5.1211
CAS-Reg.No.: 438-23-3

Androsterone **Androsteron**

= Androstan-17-one, 3α-hydroxy-(5α)-

Anethole **Anethol**

sys[CA]: Benzene, 1-methoxy-4-(1-propenyl)-

$H_3C-CH=CH-C_6H_4-O-CH_3$

$C_{10}H_{12}O$
Beilstein: 4-6.3796 · Syst.-Nr. 534
CAS-Reg.No.: 104-46-1
IUPAC: C-214.1

cis-Anethole
CAS-Reg.No.: 25679-28-1
trans-Anethole
CAS-Reg.No.: 4180-23-8

Angelic acid **Angelicasäure**

= *cis*-Crotonic acid, 2-methyl-

Aniline **Anilin**

sys[CA]: Benzenamine

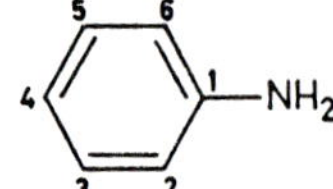

C_6H_7N
Beilstein: 4-12.223 · Syst.-Nr. 1598
CAS-Reg.No.: 62-53-3
IUPAC: C-812.1

Anisic acid **Anissäure**

sys[B]: Benzoesäure, Methoxy-
[CA]: Benzoic acid, methoxy-

$C_8H_8O_3$
CAS-Reg.No.: 1335-08-6
IUPAC: C-411.1

p-

o-Anisic acid
Beilstein: 4-10.130 · Syst.-Nr. 1059
CAS-Reg.No.: 579-75-9
m-Anisic acid
Beilstein: 4-10.316 · Syst.-Nr. 1068
CAS-Reg.No.: 586-38-9
p-Anisic acid
Beilstein: 4-10.346 · Syst.-Nr. 1069
CAS-Reg.No.: 100-09-4

Anisidine **Anisidin**

sys[CA]: Benzenamine, methoxy-

C_7H_9NO
CAS-Reg.No.: 29191-52-4
IUPAC: C-812.1

p-

o-Anisidine
Beilstein: 4-13.806 · Syst.-Nr. 1829
CAS-Reg.No.: 90-04-0
m-Anisidine
Beilstein: 4-13.953 · Syst.-Nr. 1840
CAS-Reg.No.: 536-90-3
p-Anisidine
Beilstein: 4-13.1015 · Syst.-Nr. 1843
CAS-Reg.No.: 104-94-9

Anisole **Anisol**

sys[CA]: Benzene, methoxy-

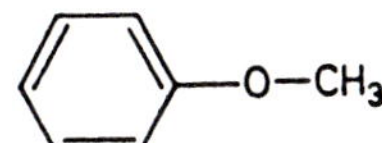

C_7H_8O
Beilstein: 4-6.548 · Syst.-Nr. 514
CAS-Reg.No.: 100-66-3
IUPAC: C-214.1

Annotinine **Annotinin**

$C_{16}H_{21}NO_3$
Beilstein: 4-27.6563 · Syst.-Nr. 4444
CAS-Reg.No.: 559-49-9

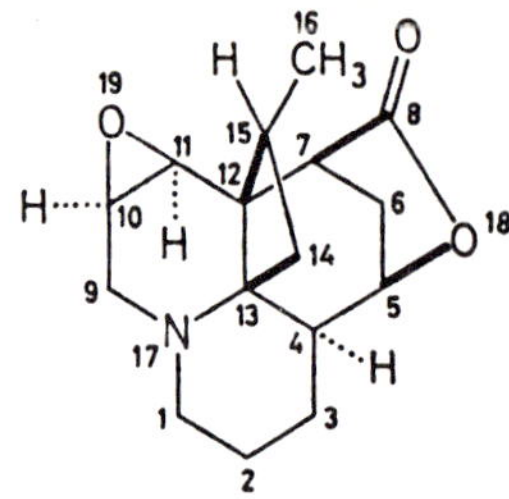

[n]Annulene **[n]Annulen**

Collective name for Cycloalkapolyenes; the number in square brackets before the name designates the number of C-atoms in the ring.

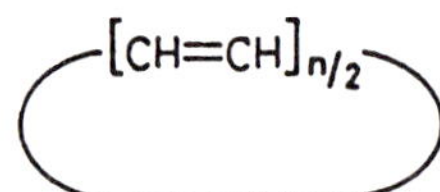

Anthracene **Anthracen**

$C_{14}H_{10}$
Beilstein: 4-5.2281 · Syst.-Nr. 482
CAS-Reg.No.: 120-12-7
IUPAC: A-21.2(13)

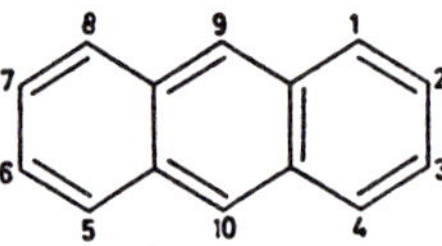

Anthranilic acid **Anthranilsäure**

sys[CA]: Benzoic acid, 2-amino-

$C_7H_7NO_2$
Beilstein: 4-14.1004 · Syst.-Nr. 1889
CAS-Reg.No.: 118-92-3
IUPAC: C-421.4

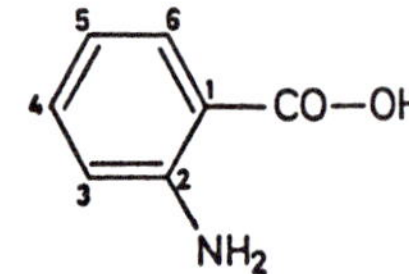

Anthrazine **Anthrazin**

sys[B]: Dinaphtho[2,3-*a*;2′,3′-*h*]phenazin

$C_{28}H_{16}N_2$
Beilstein: 4-23.2203 · Syst.-Nr. 3499
CAS-Reg.No.: 222-64-0

Anthyridine **Anthyridin**

sys[B]: Pyrido[2,3-*b*][1,8]naphthyridin

$C_{11}H_7N_3$
Beilstein · Syst.-Nr. 3812
CAS-Reg.No.: 261-15-4

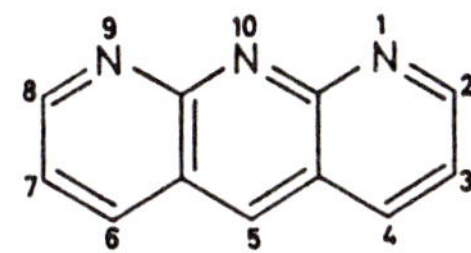

Antiarose **Antiarose**

= D-Gulose, 6-Deoxy-

Apiose **Apiose**

$C_5H_{10}O_5$
Beilstein · Syst.-Nr. 135

D-Apiose
Beilstein: 4-1.4258
CAS-Reg.No.: 639-97-4

L-Apiose
Beilstein: 4-1.4258
CAS-Reg.No.: 6477-44-7

DL-Apiose
Beilstein: 4-1.4258
CAS-Reg.No.: 42927-70-8

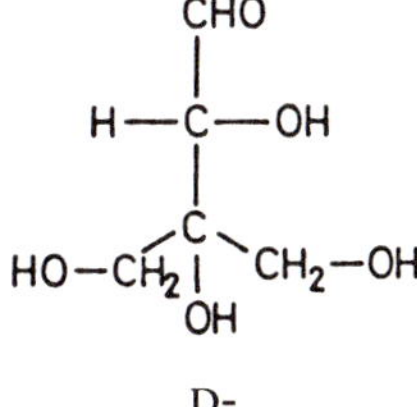
D-

Aporphane **Aporphan**

syn: Noraporphine
sys[CA]: 4*H*-Dibenzo[*de,g*]quinoline, 5,6,6a,7-tetrahydro-

$C_{16}H_{15}N$
Beilstein: 4-20.4099 · Syst.-Nr. 3088
CAS-Reg.No.: 519-01-7

–, 6-methyl- = *Aporphine*

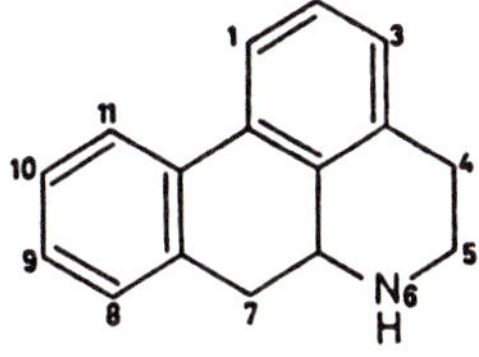

Aporphine **Aporphin**

= Aporphane, 6-methyl-

Apotrichothecane **Apotrichothecan**

$C_{15}H_{26}O$
Beilstein: 4-18.1216* · Syst.-Nr. 2364
CAS-Reg.No.: 24384-10-9
IUPAC: F-(10)

Arabinofuranose **Arabinofuranose**

$C_5H_{10}O_5$
Beilstein · Syst.-Nr. 133
CAS-Reg.No.: 13221-22-2
IUPAC: Carb-5/18

α-D-Arabinofuranose
Beilstein: 4-1.4215
CAS-Reg.No.: 37388-49-1
β-D-Arabinofuranose
Beilstein: 4-1.4215
CAS-Reg.No.: 25545-03-3
α-L-Arabinofuranose
CAS-Reg.No.: 38029-69-5
β-L-Arabinofuranose
CAS-Reg.No.: 20074-49-1

HO—$\overset{5}{C}H_2$ O OH 4 1 3 2 HO OH

β-D-

Arabinopyranose **Arabinopyranose**

$C_5H_{10}O_5$
Beilstein · Syst.-Nr. 133
IUPAC: Carb-5/18

α-D-Arabinopyranose
Beilstein: 4-1.4215
CAS-Reg.No.: 608-45-7
β-D-Arabinopyranose
Beilstein: 4-1.4215
CAS-Reg.No.: 6748-95-4
α-L-Arabinopyranose
Beilstein: 4-1.4218
CAS-Reg.No.: 7296-55-1
β-L-Arabinopyranose
Beilstein: 4-1.4218
CAS-Reg.No.: 7296-56-2

5 O HO 4 1 OH 3 2 HO OH

β-D-

Arabinose **Arabinose**

$C_5H_{10}O_5$
Beilstein · Syst.-Nr. 133
CAS-Reg.No.: 147-81-9
IUPAC: Carb-5

D-Arabinose
Beilstein: 4-1.4215
CAS-Reg.No.: 10323-20-3
L-Arabinose
Beilstein: 4-1.4217
CAS-Reg.No.: 5328-37-0
DL-Arabinose
Beilstein: 4-1.4223
CAS-Reg.No.: 20235-19-2

D-

Arcanose **Arcanose**

= *xylo*-2,6-Dideoxy-hexose,
3,O^3-dimethyl-

Arginine **Arginin**

$C_6H_{14}N_4O_2$

$HN{=}C(NH_2){-}CH_2{-}CH_2{-}CH_2{-}CH(NH_2){-}CO{-}OH$

Beilstein · Syst.-Nr. 367
CAS-Reg.No.: 7004-12-8
IUPAC: 3AA-1

D-Arginine
Beilstein: 4-4.2648
CAS-Reg.No.: 157-06-2
L-Arginine
Beilstein: 4-4.2648
CAS-Reg.No.: 74-79-3
DL-Arginine
Beilstein: 3-4.1359
CAS-Reg.No.: 7200-25-1

Arsanthrene **Arsanthren**

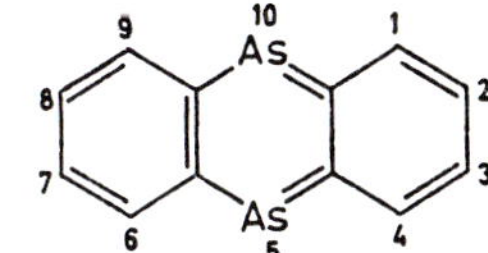

$C_{12}H_8As_2$
Beilstein · Syst.-Nr. 4720
CAS-Reg.No.: 260-22-0

Arsanthridine **Arsanthridin**

$C_{13}H_9As$
Beilstein · Syst.-Nr. 4720
CAS-Reg.No.: 229-63-0

Artostane **Artostan**

= Lanostane, (5α)-9β,19-cyclo-

Ascarylose **Ascarylose**

= L-*arabino*-3,6-Dideoxy-hexose

Ascorbic acid **Ascorbinsäure**

$C_6H_8O_6$
Beilstein · Syst.-Nr. 2568
IUPAC: M-16

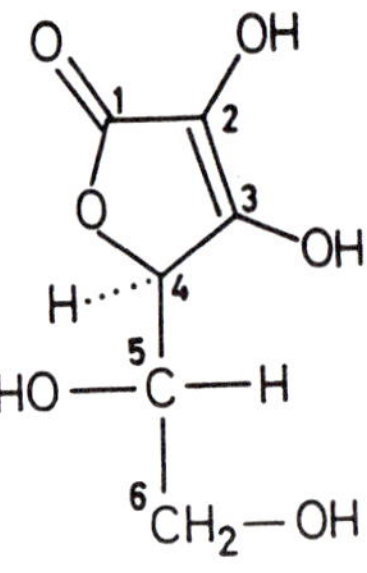

L-

D-Ascorbic acid
Beilstein: 4-18.3046
CAS-Reg.No.: 10504-35-5
L-Ascorbic acid [syn: Vitamin C]
Beilstein: 4-18.3038
CAS-Reg.No.: 50-81-7
DL-Ascorbic acid
Beilstein: 4-18.3047
CAS-Reg.No.: 62624-30-0

Asparagine **Asparagin**

$C_4H_8N_2O_3$

$H_2N-CO-CH_2-CH(NH_2)-CO-OH$

Beilstein · Syst.-Nr. 372
CAS-Reg.No.: 7006-34-0
IUPAC: 3AA-1

D-Asparagine
Beilstein: 4-4.3004
CAS-Reg.No.: 2058-58-4
L-Asparagine
Beilstein: 4-4.3005
CAS-Reg.No.: 70-47-3
DL-Asparagine
Beilstein: 4-4.3005
CAS-Reg.No.: 3130-87-8

Aspartic acid **Asparaginsäure**

$C_4H_7NO_4$

$HO-CO-CH_2-CH(NH_2)-CO-OH$

Beilstein · Syst.-Nr. 372
CAS-Reg.No.: 6899-03-2
IUPAC: 3AA-1

D-Aspartic acid
Beilstein: 4-4.2998
CAS-Reg.No.: 1783-96-6

L-Aspartic acid
Beilstein: 4-4.2998
CAS-Reg.No.: 56-84-8

DL-Aspartic acid
Beilstein: 4-4.3000
CAS-Reg.No.: 617-45-8

Aspidofractinine **Aspidofractinin**

$C_{19}H_{24}N_2$
Beilstein: 4-25.980* · Syst.-Nr. 3486
CAS-Reg.No.: 6871-25-6

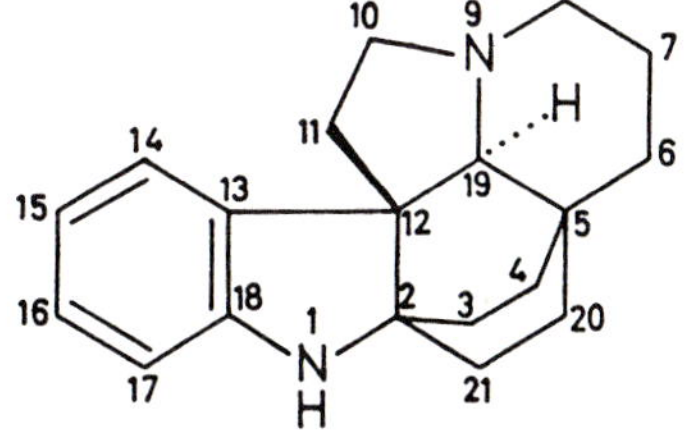

Aspidospermidine **Aspidospermidin**

$C_{19}H_{26}N_2$
Beilstein: 4-22.5702* · Syst.-Nr. 3485
CAS-Reg.No.: 2912-09-6

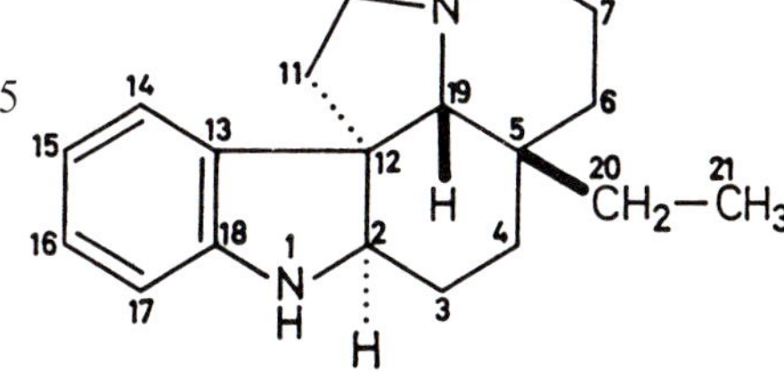

Atidane **Atidan**

$C_{19}H_{31}N$
Beilstein: 4-20.3327* · Syst.-Nr. 3065
CAS-Rcg.No.: 41904 84-1
IUPAC: F-2

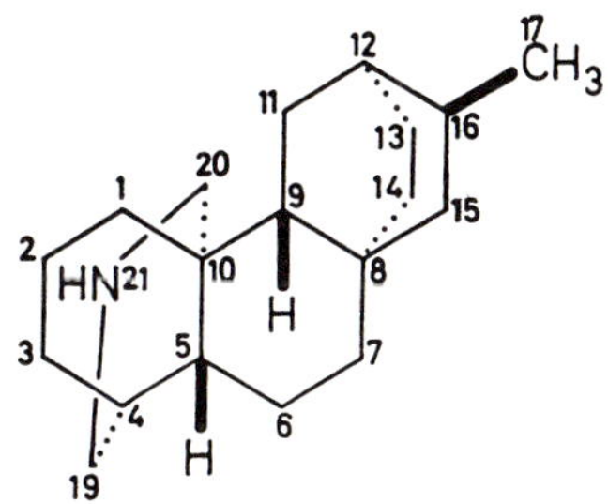

Atisane **Atisan**

$C_{20}H_{34}$
Beilstein: 4-18.6084* · Syst.-Nr. 472
CAS-Reg.No.: 24379-83-7
IUPAC: F-2

Atisine **Atisin**

$C_{22}H_{33}NO_2$
Beilstein: 4-27.1998 · Syst.-Nr. 4223
CAS-Reg.No.: 466-43-3

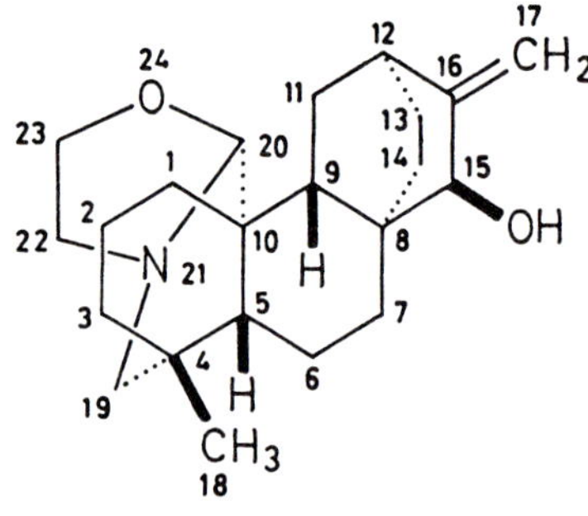

Axenose **Axenose**

= *xylo*-2,6-Dideoxy-hexose, 3-methyl-

Axerol **Axerol**

= Retinol

Axerophthene **Axerophthen**

= Deoxyretinol

Axerophthol **Axerophthol**

= Retinol

Azelaic acid **Azelainsäure**

= Nonanedioic acid

Azulene **Azulen**

$C_{10}H_8$
Beilstein: 4-5.1636 · Syst.-Nr. 475 A
CAS-Reg.No.: 275-51-4
IUPAC: A-21.2(4)

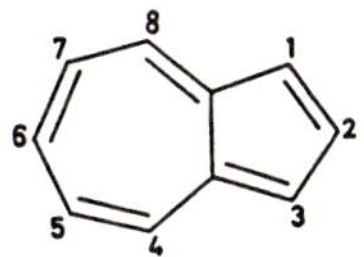

–, (3a*S*)-5*c*-isopropyl-3a,8*c*-dimethyl-(3a*r*,8a*t*)-decahydro- = Ambrosane
–, (3a*S*)-7*t*-isopropyl-1*t*,4*t*-dimethyl-(3a*r*,8a*c*)-decahydro- = Guaiane

Bacteriochlorin **Bacteriochlorin**

sys[B]: Porphyrin, 2,3,12,13-Tetrahydro-
[CA]: 21*H*,23*H*-Porphine, 7,8,17,18-tetrahydro-

$C_{20}H_{18}N_4$
Beilstein · Syst.-Nr. 4029
CAS-Reg.No.: 2683-78-5
IUPAC: TP-4.1

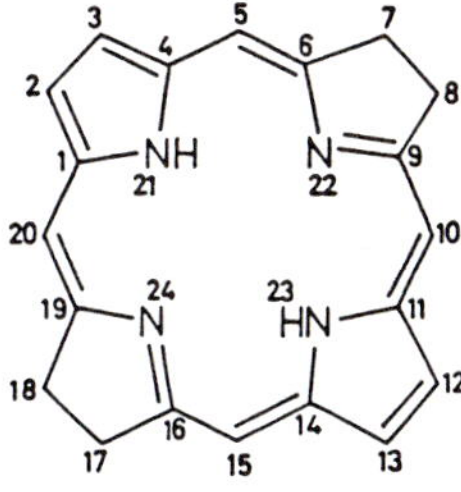

Bakkane **Bakkan**

sys[CA]: 1*H*-Indene, octahydro-2,3a,4-trimethyl-2-(1-methylethyl)-, [2*S*-(2α,3aβ,4β,8aβ)]-

$C_{15}H_{28}$
Beilstein · Syst.-Nr. 453
IUPAC: F-2

Barbituric acid **Barbitursäure**

sys[CA]: 2,4,6(1*H*,3*H*,5*H*)-Pyrimidinetrione

$C_4H_4N_2O_3$
Beilstein: 4-24.1873 · Syst.-Nr. 3615
CAS-Reg.No.: 67-52-7

Benzene **Benzol**

C_6H_6
Beilstein: 4-5.583 · Syst.-Nr. 463
CAS-Reg.No.: 71-43-2
IUPAC: A-11.3

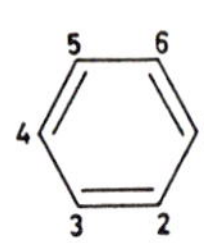

–, isopropyl- = Cumene
–, isopropyl-methyl- = Cymene
–, methoxy- = Anisole
–, 1-methoxy-4-propenyl- = Anethole

Benzidine **Benzidin**

sys[CA]: [1,1′-Biphenyl]-4,4′-diamine

$C_{12}H_{12}N_2$
Beilstein: 4-13.364 · Syst.-Nr. 1786
CAS-Reg.No.: 92-87-5
IUPAC: C-813.1

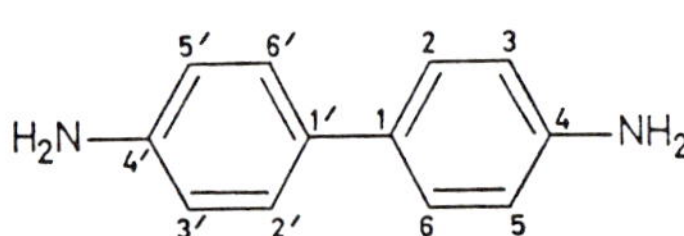

Benzil **Benzil**

sys[CA]: Ethanedione, diphenyl-

$C_{14}H_{10}O_2$
Beilstein: 4-7.2502 · Syst.-Nr. 677
CAS-Reg.No.: 134-81-6
IUPAC: C-313.4

Benzilic acid **Benzilsäure**

sys[B]: Essigsäure, Hydroxy-diphenyl-
[CA]: Benzeneacetic acid,
α-hydroxy-α-phenyl-

$C_{14}H_{12}O_3$
Beilstein: 4-10.1256 · Syst.-Nr. 1089
CAS-Reg.No.: 76-93-7
IUPAC: C-411.1

Benzoic acid **Benzoesäure**

$C_7H_6O_2$
Beilstein: 4-9.273 · Syst.Nr. 897
CAS-Reg.No.: 65-85-0
IUPAC: C-404.1

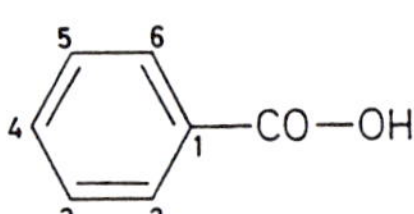

–, 2-amino- = Anthranilic acid
–, methoxy- = Anisic acid

Benzol = Benzene

Benzyl alcohol **Benzylalkohol**

sys[CA]: Benzenemethanol

C_7H_8O
Beilstein: 4-6.2222 · Syst.-Nr. 528 A
CAS-Reg.No.: 100-51-6
IUPAC: C-201.4

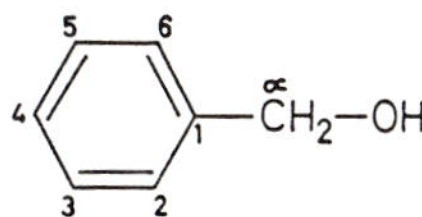

Berbaman **Berbaman**

$C_{32}H_{30}N_2O_2$
Beilstein: 4-19.4279* · Syst.-Nr. 4635
CAS-Reg.No.: 34159-88-1
IUPAC: F-2

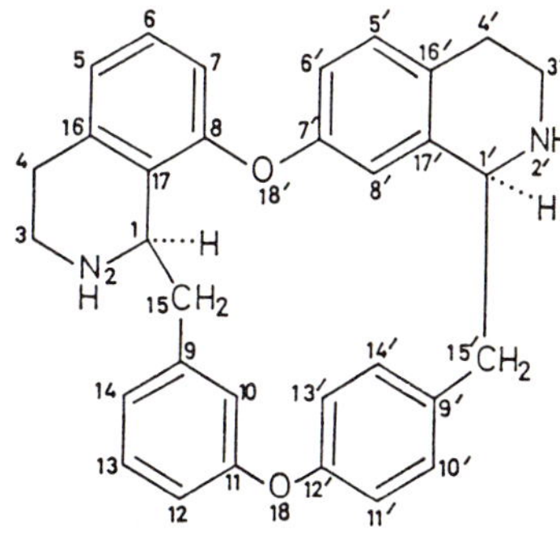

Berbine **Berbin**

sys[CA]: 6H-Dibenzo[*a,g*]quinolizine, 5,8,13,13a-tetrahydro-

$C_{17}H_{17}N$
Beilstein: 4-20.4108 · Syst.-Nr. 3088
CAS-Reg.No.: 483-49-8

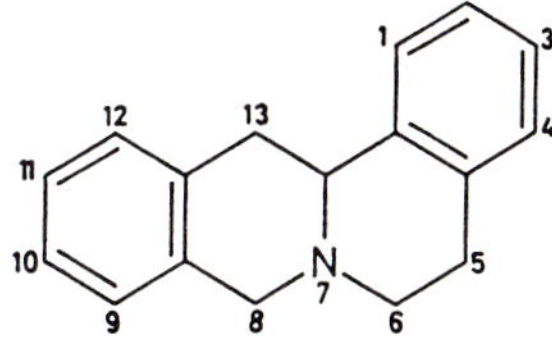

Bernsteinsäure = Succinic acid

Betaine **Betain**

sys[B]: Ammonium, Carboxymethyl-trimethyl-, betain
[CA]: Methanaminium, 1-carboxy-*N,N,N*-trimethyl-, hydroxide, inner salt

$C_5H_{11}NO_2$
Beilstein: 4-4.2369 · Syst.-Nr. 364
CAS-Reg.No.: 107-43-7
IUPAC: C-816.1

$(CH_3)_3N^+\text{-}CH_2\text{-}CO\text{-}O^-$

Beyerane **Beyeran**

= Phyllocladane, 13-methyl-17-nor-

Bilin **Bilin**

sys[CA]: 21*H*-Biline

$C_{19}H_{14}N_4$
Beilstein: 4-26.1894* · Syst.-Nr. 4030
CAS-Reg.No.: 24755-35-9
IUPAC: TP-6.1

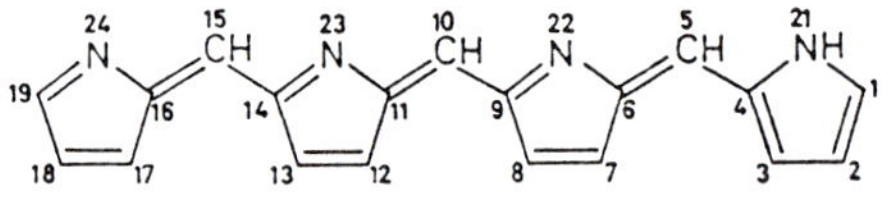

–, 5,10,15,21,23,24-hexahydro-
= *Bilinogen*

Bilinogen **Bilinogen**

= Bilin, 5,10,15,21,23,24-hexahydro-

Bilirubin **Bilirubin**

syn: Bilirubin IXα
sys[CA]: 21*H*-Biline-8,12-dipropanoic acid, 2,17-diethenyl-1,10,19,22,23,24-hexahydro-3,7,13,18-tetramethyl-1,19-dioxo-

$C_{33}H_{36}N_4O_6$
Beilstein: 4-26.3268 · Syst.-Nr. 4173
CAS-Reg.No.: 635-65-4
IUPAC: TP-6.4

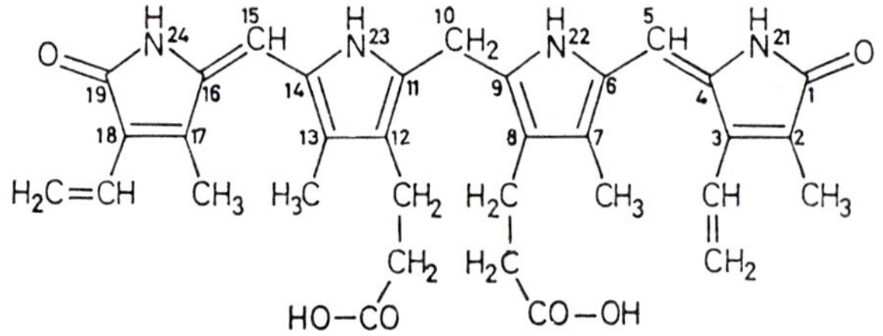

Biliverdin **Biliverdin**

syn: Biliverdin IXd

sys[CA]: 21*H*-Biline-8,12-dipropanoic acid, 3,18-diethenyl-1,19,22,24-tetrahydro-2,7,13,17-tetramethyl-1,19-dioxo-

$C_{33}H_{34}N_4O_6$
Beilstein: 4-26.3272 · Syst.-Nr. 4173
CAS-Reg.No.: 114-25-0
IUPAC: TP-6.4

–, 10,23-dihydro- = Bilirubin

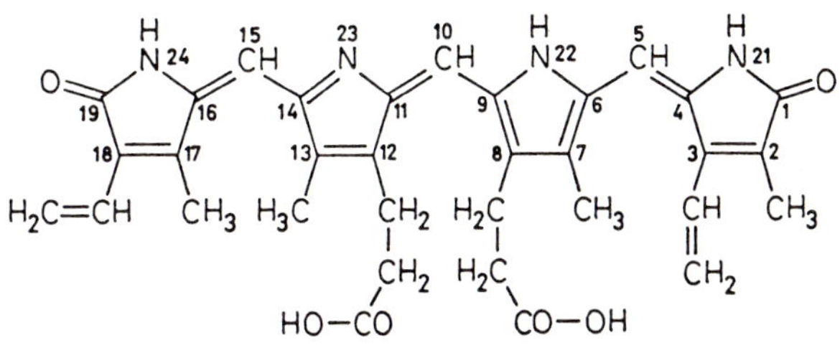

Biphenyl **Biphenyl**

$C_{12}H_{10}$
Beilstein: 4-5.1807 · Syst.-Nr. 479
CAS-Reg.No.: 92-52-4
IUPAC: A-52.4

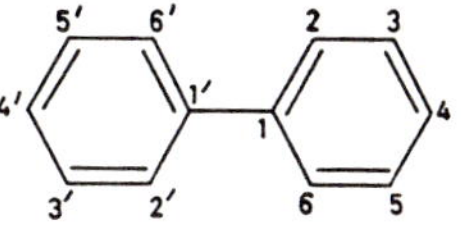

Biphenylene **Biphenylen**

$C_{12}H_8$
Beilstein: 4-5.2137 · Syst.-Nr. 480
CAS-Reg.No.: 259-79-0
IUPAC: A-21.2(6)

Biuret **Biuret**

sys[CA]: Imidodicarbonic diamide

$C_2H_5N_3O_2$
Beilstein: 4-3.141 · Syst.-Nr. 205 A
CAS-Reg.No.: 108-19-0
IUPAC: C-975.1

$H_2N-CO-NH-CO-NH_2$

Boivinose **Boivinose**

= *xylo*-2,6-Dideoxy-hexose

Bornane **Bornan**

sys[CA]: Bicyclo[2.2.1]heptane, 1,7,7-trimethyl-

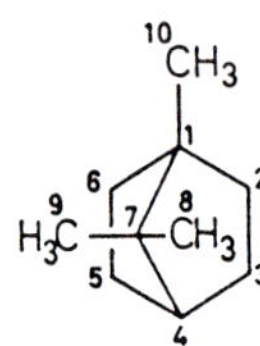

$C_{10}H_{18}$
Beilstein: 4-5.319 · Syst.-Nr. 453
CAS-Reg.No.: 464-15-3
IUPAC: A-72.1

Brenzcatechin = Pyrocatechol

Brenztraubensäure = Pyruvic acid

Bufalin **Bufalin**

= Bufa-20,22-dienolide, 3β,14-dihydroxy-(5β,14β)-

Bufanolide **Bufanolid**

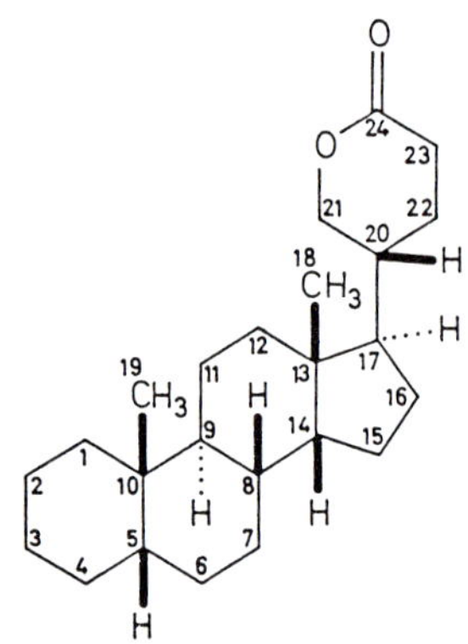

$C_{24}H_{38}O_2$
Beilstein: 4-17.5042 · Syst.-Nr. 2463
CAS-Reg.No.: 24887-77-2
IUPAC: 2S-3.2(43)

Bullvalene **Bullvalen**

sys[CA]: Tricyclo[3.3.2.$0^{2,8}$]deca-3,6,9-triene

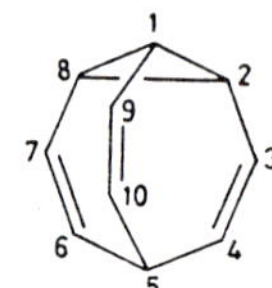

$C_{10}H_{10}$
Beilstein · Syst.-Nr. 474
CAS-Reg.No.: 1005-51-2

Butane **Butan**

C_4H_{10}
Beilstein: 4-1.236 · Syst.-Nr. 10
CAS-Reg.No.: 106-97-8
IUPAC: A-1.1

$CH_3-CH_2-CH_2-CH_3$

Buttersäure = Butyric acid

Butyraldehyde **Butyraldehyd**
sys[CA]: Butanal

C_4H_8O
Beilstein: 4-1.3229 · Syst.-Nr. 87
CAS-Reg.No.: 123-72-8
IUPAC: C-305.1

$CH_3-CH_2-CH_2-CHO$

Butyric acid **Buttersäure**
sys[CA]: Butanoic acid

$C_4H_8O_2$
Beilstein: 4-2.779 · Syst.-Nr. 162
CAS-Reg.No.: 107-92-6
IUPAC: C-404.1

$CH_3-CH_2-CH_2-CO-OH$

–, 3-methyl- = *Isovaleric acid*
–, 3-oxo- = Acetoacetic acid

Cadinane **Cadinan**
sys[CA]: Naphthalene, decahydro-1,6-dimethyl-4-(1-methylethyl)-, [1*S*-(1α,4α,4aα,6α,8aβ)]-

$C_{15}H_{28}$
Beilstein: 4-5.354 · Syst.-Nr. 453
CAS-Reg.No.: 483-73-8

Calciol — **Calciol**

syn: Cholecalciferol
sys[CA]: 9,10-Secocholesta-5,7,10(19)-trien-3-ol, (3β,5*Z*,7*E*)-

$C_{27}H_{44}O$
Beilstein: 4-6.4149 · Syst.-Nr. 535
CAS-Reg.No.: 67-97-0
IUPAC: VD-3

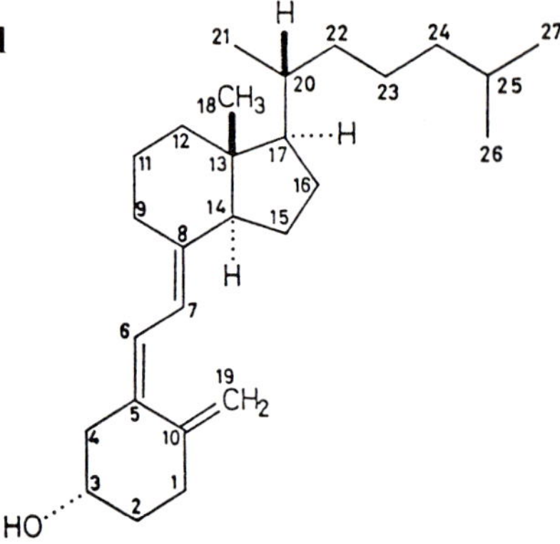

C-**Calebassine** — *C*-**Calebassin**

$[C_{40}H_{48}N_4O_2]^{2+}$
Beilstein: 4-26.2040 · Syst.-Nr. 4082
CAS-Reg.No.: 7257-29-6

Calycanthidine — **Calycanthidin**

$C_{23}H_{28}N_4$
Beilstein: 4-26.1865 · Syst.-Nr. 4027
CAS-Reg.No.: 5516-85-8

Calycanthine **Calycanthin**

$C_{22}H_{26}N_4$
Beilstein: 4-26.1863 · Syst.-Nr. 4027
CAS-Reg.No.: 595-05-1

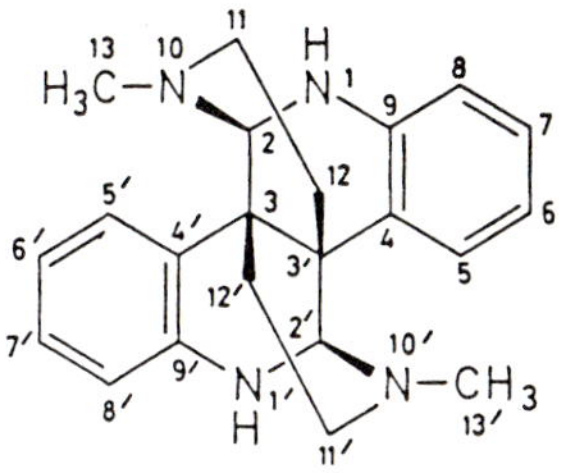

Camphane **Camphan**

= Bornane

Camphene **Camphen**

sys[CA]: Bicyclo[2.2.1]heptane, 2,2-dimethyl-3-methylene-

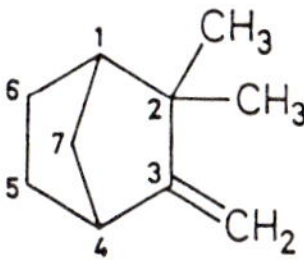

$C_{10}H_{16}$
Beilstein: 4-5.461 · Syst.-Nr. 458
CAS-Reg.No.: 79-92-5
IUPAC: A-74.3

Camphoric acid **Camphersäure**

sys[CA]: 1,3-Cyclopentanedicarboxylic acid, 1,2,2-trimethyl-

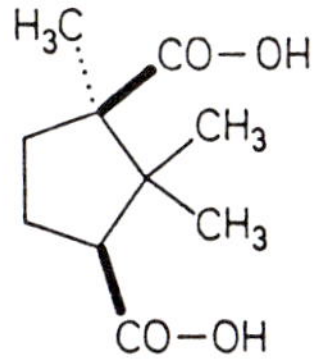

$C_{10}H_{16}O_4$
Beilstein · Syst.-Nr. 965
IUPAC: C-404.1

(1 *R*)-*cis*-

(1 *R*)-*cis*-Camphoric acid
Beilstein: 4-9.2851
CAS-Reg.No.: 124-83-4

(1 *S*)-*cis*-Camphoric acid
Beilstein: 3-9.3878

(±)-*cis*-Camphoric acid
Beilstein: 3-9.3878
CAS-Reg.No.: 560-05-4

(1 *R*)-*trans*-Camphoric acid
Beilstein: 3-9.3878
CAS-Reg.No.: 595-32-4

(±)-*trans*-Camphoric acid
Beilstein: 3-9.3879
CAS-Reg.No.: 29607-01-0

Cancentrine **Cancentrin**

$C_{36}H_{34}N_2O_7$
Beilstein: 4-27.8907 · Syst.-Nr. 4642
CAS-Reg.No.: 29477-90-5

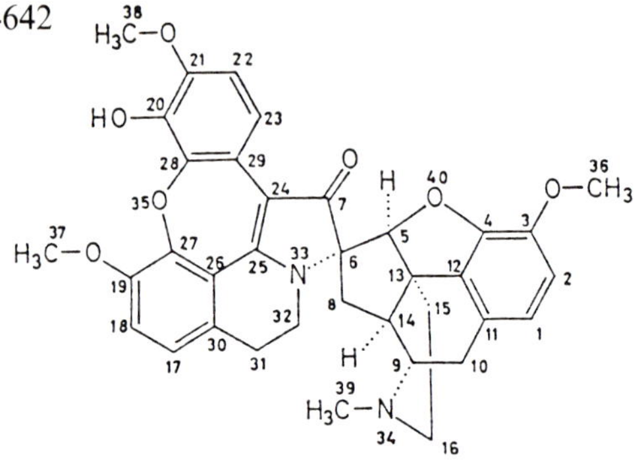

Capric acid **Caprinsäure**

= Decanoic acid

Caproic acid **Capronsäure**

= Hexanoic acid

ε-Caprolactam **ε-Caprolactam**

sys[B]: Azepin-2-on, Hexahydro-
[CA]: 2*H*-Azepin-2-one, hexahydro-

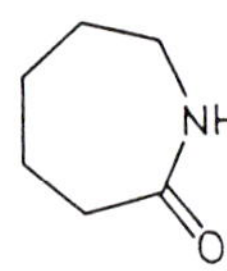

$C_6H_{11}NO$
Beilstein: 4-21.3196 · Syst.-Nr. 3179
CAS-Reg.No.: 105-60-2
IUPAC: C-475.1

Caprylic acid **Caprylsäure**

= Octanoic acid

Caracurine II **Caracurin-II**

$C_{38}H_{38}N_4O_2$
Beilstein: 4-27.9659 · Syst.-Nr. 4707
CAS-Reg.No.: 5516-84-7

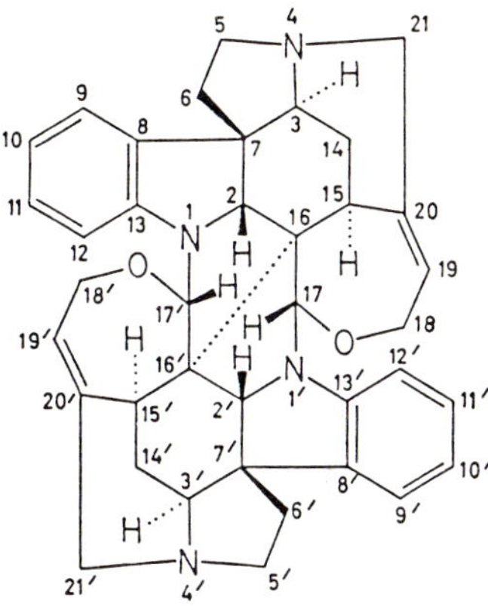

Carane **Caran**

sys[CA]: Bicyclo[4.1.0]heptane, 3,7,7-trimethyl-

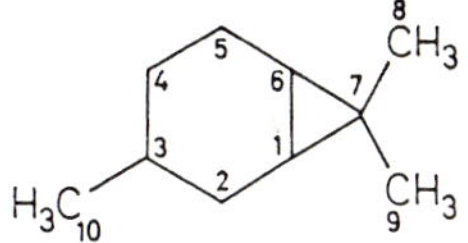

$C_{10}H_{18}$
Beilstein: 4-5.316 · Syst.-Nr. 453
CAS-Reg.No.: 554-59-6
IUPAC: A-72.1

Carbamic acid **Carbamidsäure**

CH_3NO_2
Beilstein: 4-3.37 · Syst.-Nr. 201
CAS-Reg.No.: 463-77-4
IUPAC: C-431.1

$H_2N-CO-OH$

Carbazic acid **Carbazidsäure**

sys[CA]: Hydrazinecarboxylic acid

$CH_4N_2O_2$
Beilstein: 4-3.173 · Syst.-Nr. 209 A
CAS-Reg.No.: 471-31-8
IUPAC: C-984.1

$H_2N-NH-CO-OH$

Carbazole **Carbazol**

sys[CA]: 9*H*-Carbazole

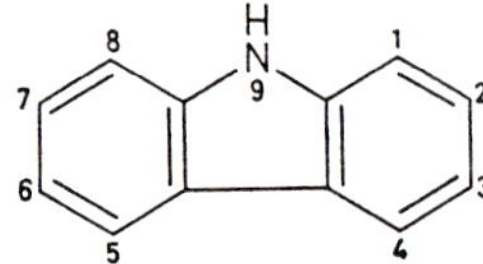

$C_{12}H_9N$
Beilstein: 4-20.3824 · Syst.-Nr. 3086
CAS-Reg.No.: 86-74-8
IUPAC: B-2.11(35)

1*H*-Carbazole
CAS-Reg.No.: 244-57-5

Carbodiimide **Carbodiimid**

sys[CA]: Methanediimine

CH_2N_2
Beilstein · Syst.-Nr. 206
CAS-Reg.No.: 151-51-9
IUPAC: C-956.1

HN=C=NH

α-Carboline **α-Carbolin**

= 9*H*-Pyrido[2,3-*b*]indole

β-Carboline **β-Carbolin**

sys[CA]: 9*H*-Pyrido[3,4-*b*]indole

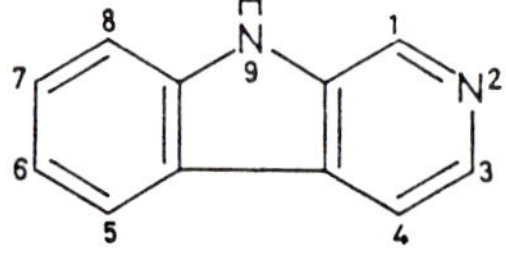

$C_{11}H_8N_2$
Beilstein: 4-23.1575 · Syst.-Nr. 3486
CAS-Reg.No.: 244-63-3
IUPAC: B-2.11(36)

γ-Carboline **γ-Carbolin**

= 5*H*-Pyrido[4,3-*b*]indole

δ-Carboline **δ-Carbolin**

= 5*H*-Pyrido[3,2-*b*]indole

Carbonic acid **Kohlensäure**

CH_2O_3
Beilstein: Derivatives of Carbonic acid
s. Syst.-Nr. 195–219 A
CAS-Reg.No.: 463-79-6

HO–CO–OH

Cardanolide **Cardanolid**

$C_{23}H_{36}O_2$
Beilstein: 4-17.5041 · Syst.-Nr. 2463
CAS-Reg.No.: 4427-84-3
IUPAC: 2S-3.1(37)

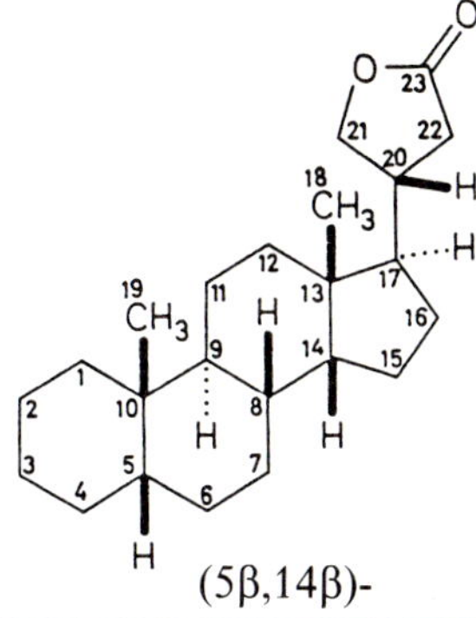

(5β,14β)-

α-Carotene **α-Carotin**

= β,ε-Carotene

β-Carotene **β-Carotin**

= β,β-Carotene

β,β-Carotene **β,β-Carotin**
syn: β-Carotene

$C_{40}H_{56}$
Beilstein: 4-5.2617 · Syst.-Nr. 488
CAS-Reg.No.: 7235-40-7
IUPAC: Car-3.2

β,ε-Carotene **β,ε-Carotin**
syn: α-Carotene

$C_{40}H_{56}$
Beilstein: 4-5.2620 · Syst.-Nr. 488
CAS-Reg.No.: 432-70-2
IUPAC: Car-3.2

β,κ-Carotene **β,κ-Carotin**

$C_{40}H_{58}$
Beilstein · Syst.-Nr. 487
CAS-Reg.No.: 36505-49-4
IUPAC: Car-3.2

β,φ-Carotene **β,φ-Carotin**

$C_{40}H_{52}$
Beilstein · Syst.-Nr. 490
CAS-Reg.No.: 4382-02-9
IUPAC: Car-3.2

β,χ-Carotene **β,χ-Carotin**

$C_{40}H_{52}$
Beilstein · Syst.-Nr. 490
CAS-Reg.No.: 55781-80-1
IUPAC: Car-3.2

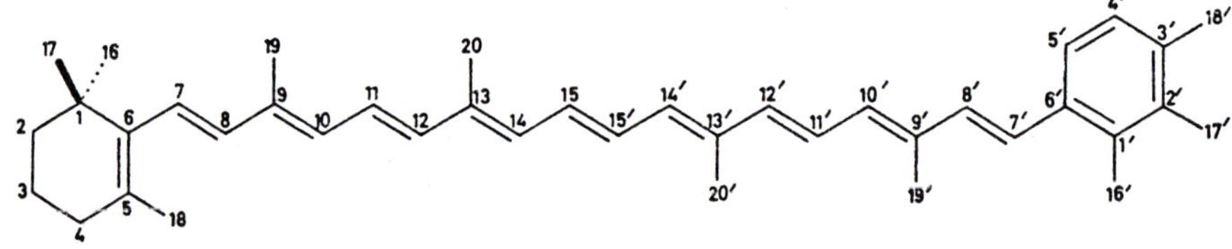

β,ψ-Carotene **β,ψ-Carotin**
syn: γ-Carotene

$C_{40}H_{56}$
Beilstein: 4-5.2616 · Syst.-Nr. 488
CAS-Reg.No.: 472-93-5
IUPAC: Car-3.2

γ-Carotene **γ-Carotin**

= β,ψ-Carotene

δ-Carotene **δ-Carotin**

= ε,ψ-Carotene

ε-Carotene **ε-Carotin**

= ε,ε-Carotene

ε,ε-Carotene **ε,ε-Carotin**

syn: ε-Carotene

$C_{40}H_{56}$
Beilstein: 4-5.2622 · Syst.-Nr. 488
CAS-Reg.No.: 38894-81-4
IUPAC: Car-3.2

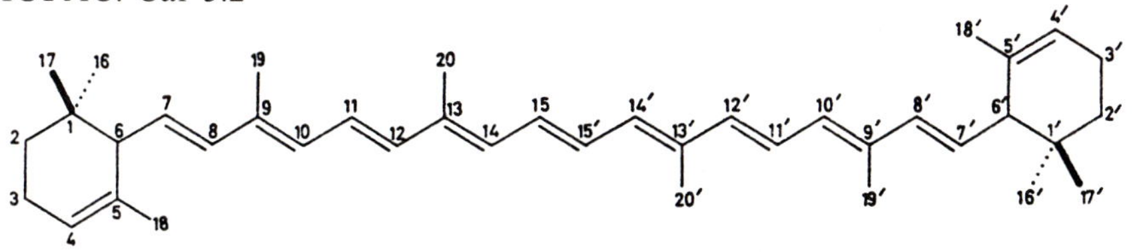

ε,ψ-Carotene **ε,ψ-Carotin**

syn: δ-Carotene

$C_{40}H_{56}$
Beilstein: 4-5.2617 · Syst.-Nr. 488
CAS-Reg.No.: 472-92-4
IUPAC: Car-3.2

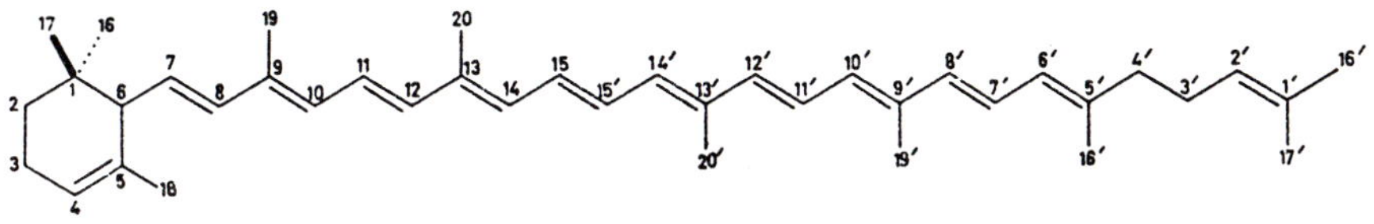

ζ-Carotene **ζ-Carotin**

= ψ,ψ-Carotene, 7,8,7′,8′-Tetrahydro-

η-Carotene **η-Carotin**

= β,β-Carotene, 7,7′,8,8′-Tetrahydro-

κ,κ-Carotene **κ,κ-Carotin**

$C_{40}H_{60}$
Beilstein · Syst.-Nr. 486
CAS-Reg.No.: 36712-49-9

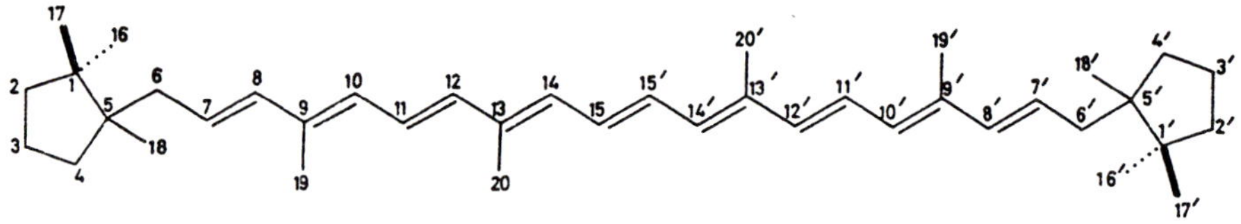

κ,φ-Carotene **κ,φ-Carotin**

$C_{40}H_{54}$
Beilstein · Syst.-Nr. 489
CAS-Reg.No.: 76684-08-7
IUPAC: Car-3.2

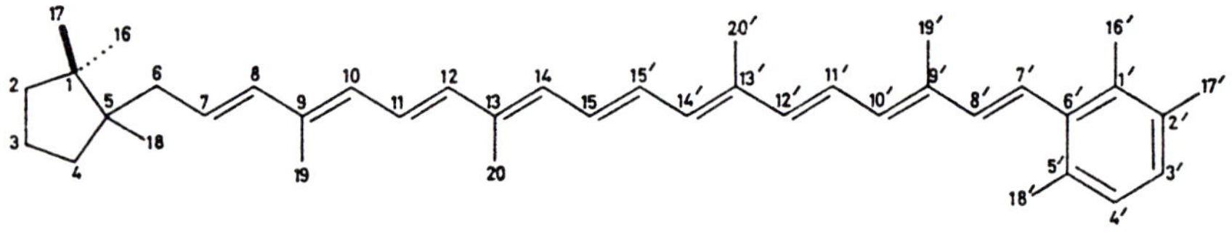

κ,χ-Carotene **κ,χ-Carotin**

$C_{40}H_{54}$
Beilstein · Syst.-Nr. 489
CAS-Reg.No.: 56144-77-5
IUPAC: Car-3.2

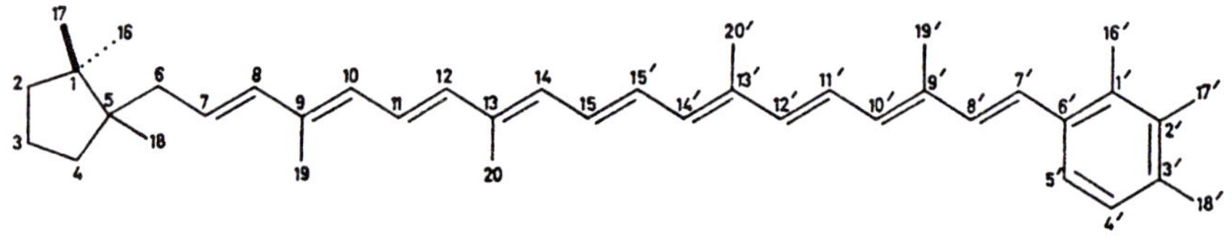

φ,φ-Carotene **φ,φ-Carotin**

$C_{40}H_{48}$
Beilstein: 4-5.2798 · Syst.-Nr. 492
CAS-Reg.No.: 524-01-6
IUPAC: Car-3.2

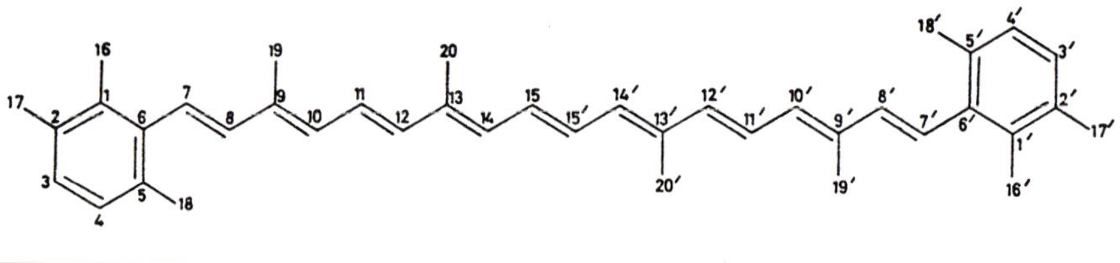

φ,χ-Carotene **φ,χ-Carotin**

$C_{40}H_{48}$
Beilstein: 4-5.2798 · Syst.-Nr. 492
CAS-Reg.No.: 550-29-8
IUPAC: Car-3.2

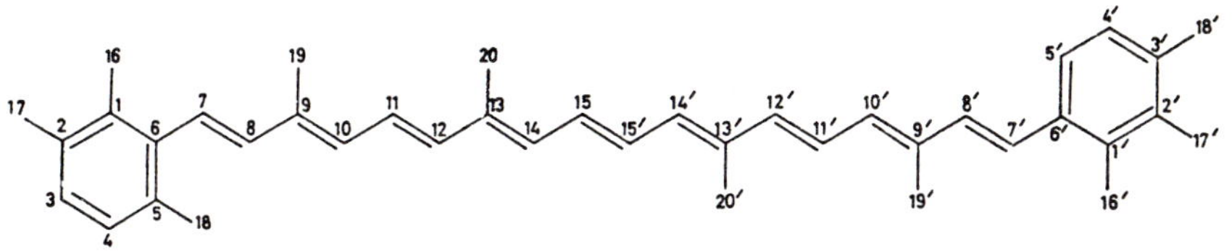

φ,ψ-Carotene **φ,ψ-Carotin**

$C_{40}H_{52}$
Beilstein · Syst.-Nr. 490
CAS-Reg.No.: 2932-09-4
IUPAC: Car-3.2

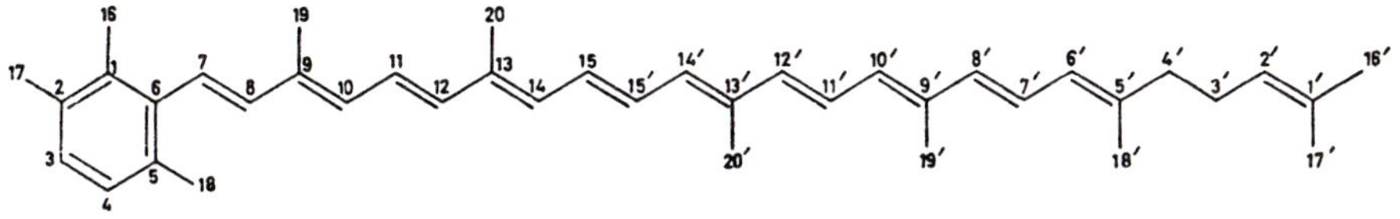

χ,χ-Carotene **χ,χ-Carotin**

$C_{40}H_{48}$
Beilstein: 4-5.2797 · Syst.-Nr. 492
CAS-Reg.No.: 6805-08-9
IUPAC: Car-3.2

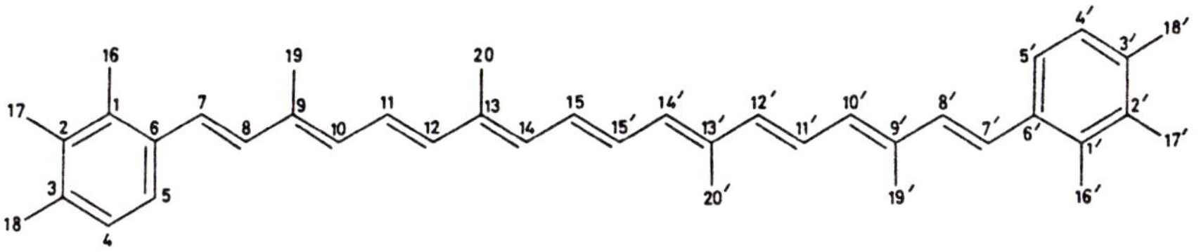

ψ,ψ-Carotene **ψ,ψ-Carotin**

$C_{40}H_{56}$
Beilstein: 4-1.1166 · Syst.-Nr. 15
CAS-Reg.No.: 502-65-8
IUPAC: Car-3.2

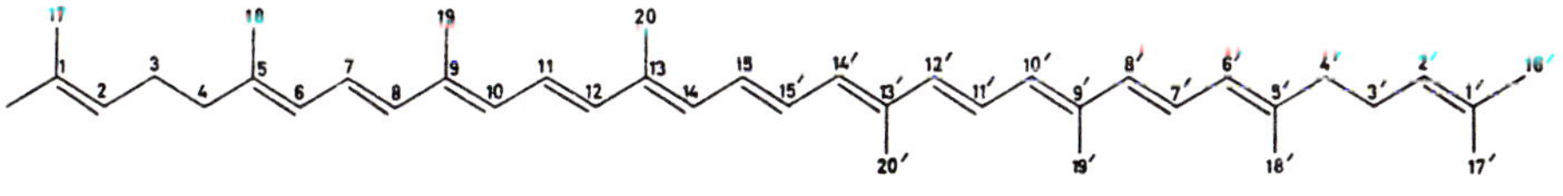

Carpaine **Carpain**

$C_{28}H_{50}N_2O_4$
Beilstein: 4-27.8782 · Syst.-Nr. 4641
CAS-Reg.No.: 3463-92-1

Carvacrol **Carvacrol**

sys[CA]: Phenol, 2-methyl-5-(1-methylethyl)-

$C_{10}H_{14}O$
Beilstein: 4-6.3331 · Syst.-Nr. 531
CAS-Reg.No.: 499-75-2
IUPAC: C-202.2

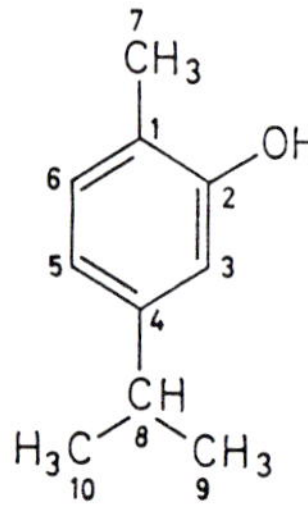

Caryolane **Caryolan**

sys[CA]: Tricyclo[6.3.1.0^{2,5}]dodecane, 4,4,8-trimethyl-, (1α,2β,5α,8α)-

$C_{15}H_{26}$
Beilstein: 3-6.4180* · Syst.-Nr. 461
CAS-Reg.No.: 484-86-6
IUPAC: F-2

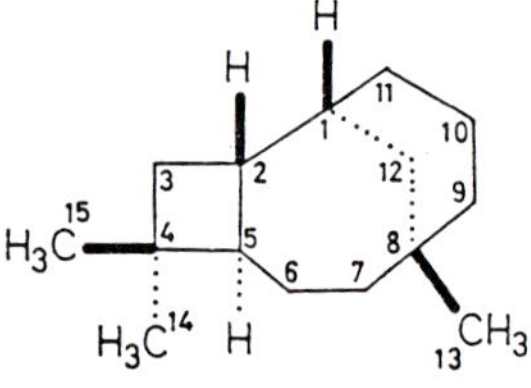

Caryoptinol **Caryoptinol**

$C_{24}H_{34}O_8$
Beilstein · Syst.-Nr. 2956
CAS-Reg.No.: 53683-47-9

Cassane **Cassan**

= Pimarane, 14-methyl-17-nor-

Cedrane **Cedran**

sys[CA]: 1*H*-3a,7-Methanoazulene, octahydro-3,6,8,8-tetramethyl-, [3*R*-(3α,3aβ,6α,7β,8aα)]-

$C_{15}H_{26}$
Beilstein: 4-18.6310* · Syst.-Nr. 461
CAS-Reg.No.: 13567-54-9
IUPAC: F-2

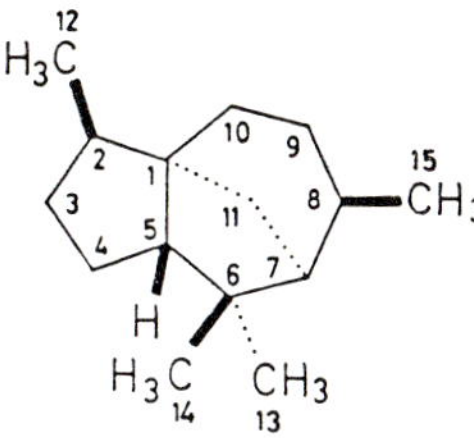

Cellobiose **Cellobiose**

= D-Glucose, O^4-β-D-glucopyranosyl-

Cembrane **Cembran**

= Cyclotetradecane, 4-isopropyl-1,7,11-trimethyl-

Cevane **Cevan**

$C_{27}H_{45}N$
Beilstein: 4-20.3331 · Syst.-Nr. 3073
CAS-Reg.No.: 482-75-7
IUPAC: S2-11(104)/F-2

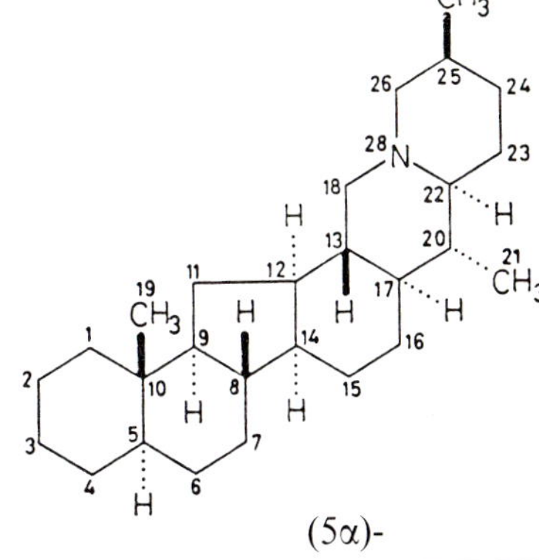

(5α)-

Cevanine **Cevanin**

–, (22*S*,25*S*)- = Cevane

Chalcone **Chalkon**

sys[CA]: 2-Propen-1-one, 1,3-diphenyl-

$C_{15}H_{12}O$
Beilstein: 4-7.1658 · Syst.-Nr. 654
CAS-Reg.No.: 94-41-7
IUPAC: C-313.2

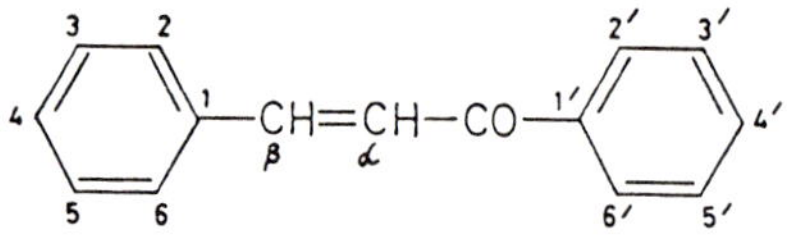

Chelidonine **Chelidonin**

$C_{20}H_{19}NO_5$
Beilstein · Syst.-Nr. 4482
CAS-Reg.No.: 476-32-4

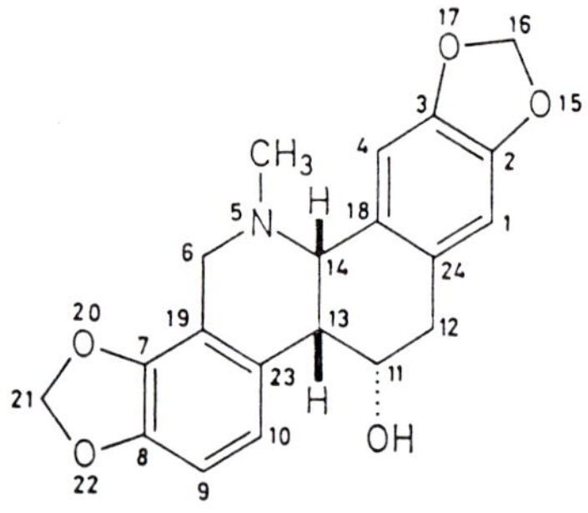

Chinasäure = Quinic acid

Chinazolin = Quinazoline

Chindolin = Quindoline

Chinindolin = Quinindoline

Chinolin = Quinoline

Chinolizidin = Quinolizidine

Chinolizin = Quinolizine

Chinovose = Quinovose

Chinoxalin = Quinoxaline

Chinucildin = Quinuclidine

Chlorin **Chlorin**

sys[B]: Porphyrin, 2,3-Dihydro-
[CA]: 21*H*,23*H*-Porphine, 2,3-dihydro-

$C_{20}H_{16}N_4$
Beilstein: 4-26.1892 · Syst.-Nr. 4030
CAS-Reg.No.: 2683-84-3
IUPAC: TP-4.1

Cholane **Cholan**

$C_{24}H_{42}$
Beilstein: 4-5.1220 · Syst.-Nr. 472
CAS-Reg.No.: 548-98-1
IUPAC: 2S-2.3(20)

(5α)-Cholane [syn: Allocholane]
CAS-Reg.No.: 6929-17-5

(5β)-

Cholanthrene **Cholanthren**

sys[CA]: Benz[*j*]aceanthrylene, 1,2-dihydro-

$C_{20}H_{14}$
Beilstein: 4-5.2638 · Syst.-Nr. 489
CAS-Reg.No.: 479-23-2
IUPAC: A-23.1

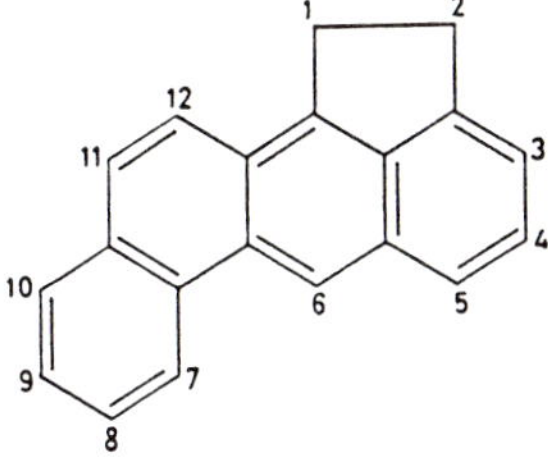

Cholecalciferol **Cholecalciferol**

= Calciol

Cholestane **Cholestan**

$C_{27}H_{48}$
Beilstein: 4-5.1226 · Syst.-Nr. 472
CAS-Reg.No.: 14982-53-7
IUPAC: 2S-2.3(20)

–, (5β)- = *Coprostane*
–, 16,23-cyclo- = Fesane
–, 12,14-cyclo-13,14-seco- = *Jervane*
–, (22*R*)-16β,22-epoxy- = Furostane
–, 4,4,14-trimethyl-(5α)- = Lanostane

(5β)-

Cholesterol **Cholesterin**

= Cholest-5-en-3β-ol

Cholic acid **Cholsäure**

= Cholan-24-oic acid,
3α,7α,12α-trihydroxy-(5β)-

Choline **Cholin**

sys[B]: Ammonium,
[2-Hydroxy-äthyl]-trimethyl-
[CA]: Ethanaminium,
2-hydroxy-*N,N,N*-trimethyl-

$[C_5H_{14}NO]^+$ $HO-CH_2-CH_2-N^+(CH_3)_3$
Beilstein: 4-4.1443 · Syst.-Nr. 353
CAS-Reg.No.: 62-49-7
IUPAC: C-816.1

Chroman **Chroman**

sys[CA]: 2*H*-1-Benzopyran,
3,4-dihydro-

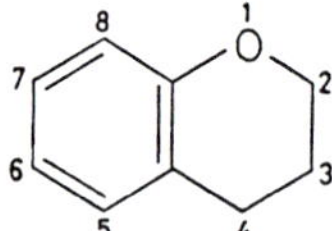

$C_9H_{10}O$
Beilstein: 5-17/1.593 · Syst.-Nr. 2366
CAS-Reg.No.: 493-08-3
IUPAC: B-2.12(2)

–, 2-phenyl- = *Flavan*
–, 3-phenyl- = *Isoflavan*

Chromene **Chromen**

sys[CA]: 2*H*-1-Benzopyran

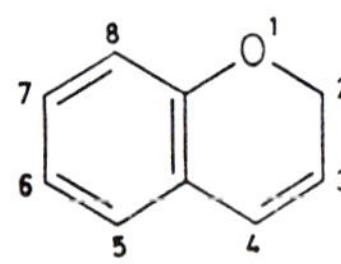

C_9H_8O
Beilstein: 5-17/2.19 · Syst.-Nr. 2367
CAS-Reg.No.: 254-04-6
IUPAC: B-2.11(8)

–, 2-phenyl- = *Flavene*
–, 3-phenyl- = *Isoflavene*

Chrysene **Chrysen**

$C_{18}H_{12}$
Beilstein: 4-5.2554 · Syst.-Nr. 488
CAS-Reg.No.: 218-01-9
IUPAC: A-21.2(19)

Cinchonan **Cinchonan**

$C_{19}H_{22}O_2$
Beilstein: 4-22.6140* · Syst.-Nr. 3487
CAS-Reg.No.: 5949-01-9

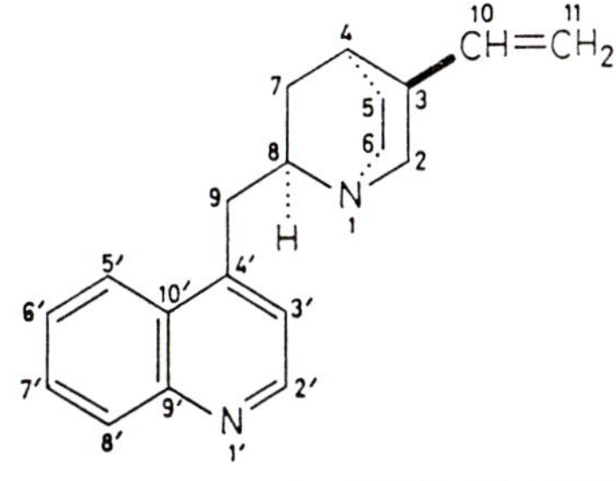

Cinnamic acid **Zimtsäure**

sys[CA]: 2-Propenoic acid, 3-phenyl-

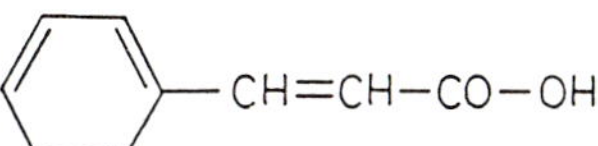

$C_9H_8O_2$
Beilstein · Syst.-Nr. 948
CAS-Reg.No.: 621-82-9
IUPAC: C-404.1

cis-Cinnamic acid
Beilstein: 4-9.2001
CAS-Reg.No.: 102-94-3

trans-Cinnamic acid
Beilstein: 4-9.2002
CAS-Reg.No.: 140-10-3

Cinnoline **Cinnolin**

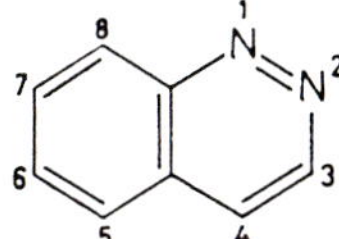

$C_8H_6N_2$
Beilstein: 4-23.1217 · Syst.-Nr. 3480
CAS-Reg.No.: 253-66-7
IUPAC: B-2.11(32)

Citraconic acid **Citraconsäure**

= Maleic acid, methyl-

Citral **Citral**

sys[B]: Octa-2,6-dienal, 3,7-Dimethyl-
[CA]: 2,6-Octadienal, 3,7-dimethyl-

$C_{10}H_{16}O$

$(CH_3)_2C=CH-CH_2-CH_2-C(CH_3)=CH-CHO$

Beilstein: 4-1.3569 · Syst.-Nr. 91
CAS-Reg.No.: 5392-40-5
IUPAC: C-305.2

Citrulline **Citrullin**

= Ornithine, N^5-carbamoyl-

Cladinose **Cladinose**

= *ribo*-2,6-Dideoxy-hexose, 3,O^3-dimethyl-

Clavularane **Clavularan**

sys[CA]: Benz[*f*]azulene, tetradecahydro-3a,5,8a-trimethyl-1-(1-methylethyl)-, (1α,3aβ,4aβ,5α,8aα,10aα)-

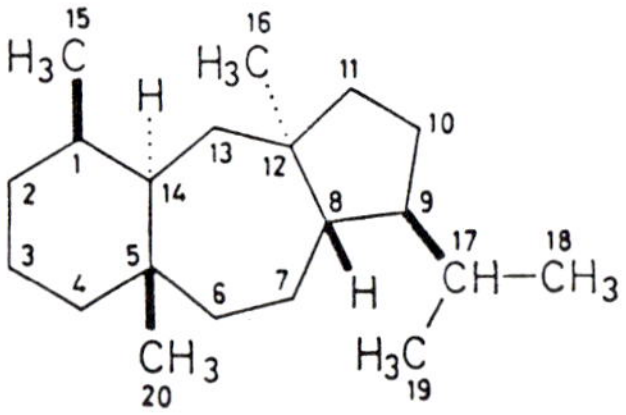

$C_{20}H_{36}$
Beilstein: 5-17/2.123* · Syst.-Nr. 461
CAS-Reg.No.: 69307-90-0
IUPAC: F-2

Clerodane **Clerodan**

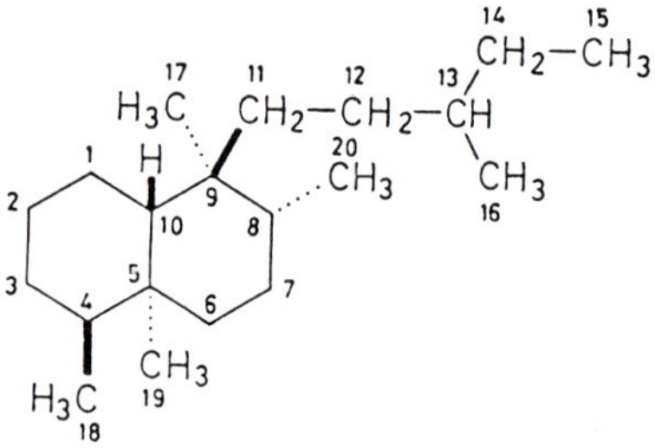

$C_{20}H_{38}$
Beilstein: 5-17/2.122* · Syst.-Nr. 453
CAS-Reg.No.: 51773-71-8
IUPAC: F-2

Clovane **Clovan**

sys[CA]: 3a,7-Methano-3a*H*-cyclopentacyclooctene, decahydro-1,1,7-trimethyl-, [3a*S*-(3aα,7α,9aβ)]-

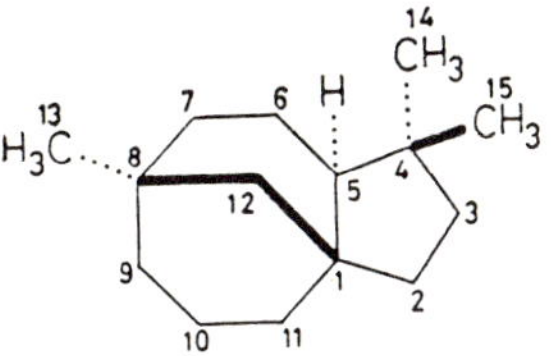

$C_{15}H_{26}$
Beilstein: 4-5.497 · Syst.-Nr. 461
CAS-Reg.No.: 469-91-0
IUPAC: F-2

Colensane **Colensan**

= Labdane, (13*S*)-8,13-epoxy-*A*-nor-

Colitose **Colitose**

= L-*xylo*-3,6-Dideoxy-hexose

Conane **Conan**

[CA]: Conanine

$C_{22}H_{37}N$
Beilstein: 4-20.3085 · Syst.-Nr. 3065
CAS-Reg.No.: 24479-76-3
IUPAC: 2S-11(100)/F-2

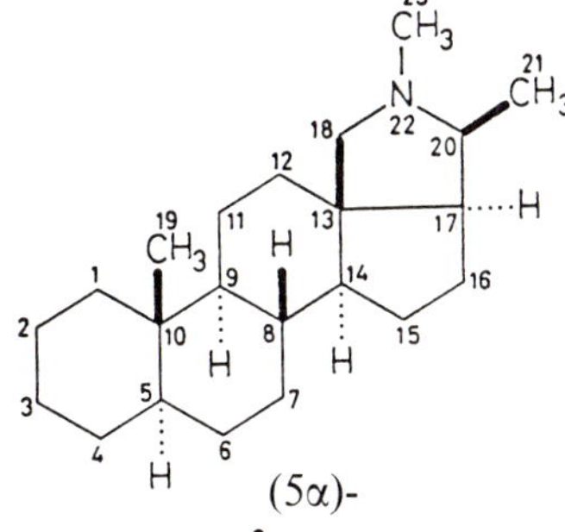

(5α)-

Conanine **Conanin**

= Conane

Condyfolan **Condyfolan**

$C_{18}H_{24}N_2$
Beilstein: 4-23.2724* · Syst.-Nr. 3485
CAS-Reg.No.: 34612-40-3
IUPAC: F-2

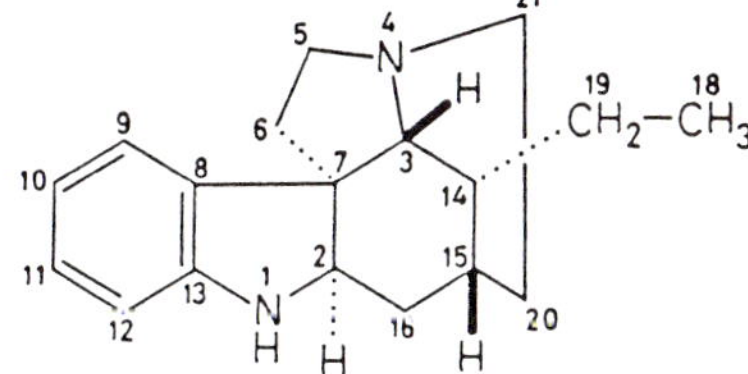

Coproporphyrin I **Koproporphyrin-I**

sys[CA]: 21*H*,23*H*-Porphine-2,7,12,17-tetrapropanoic acid, 3,8,13,18-tetramethyl-

$C_{36}H_{38}N_4O_8$
Beilstein: 4-26.3093
· Syst.-Nr. 4173
CAS-Reg.No.: 531-14-6
IUPAC: TP-2.1

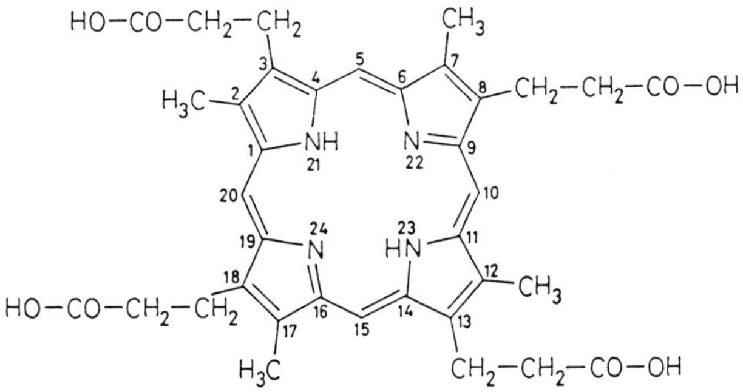

Coprostane **Koprostan**

= Cholestane, (5β)-

Coronene **Coronen**

$C_{24}H_{12}$
Beilstein: 4-5.2830 · Syst.-Nr. 494
CAS-Reg.No.: 191-07-1
IUPAC: A-21.2(30)

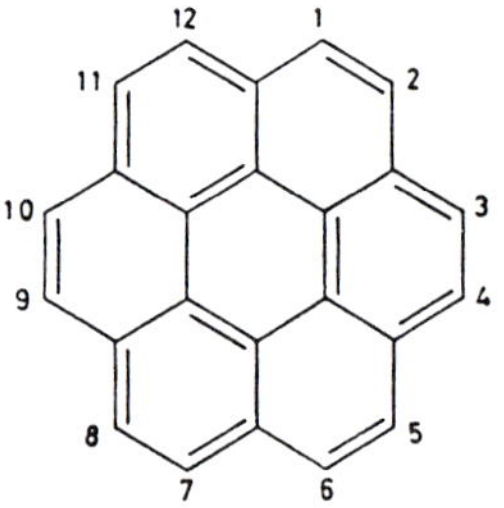

Corrin **Corrin**

$C_{19}H_{22}N_4$
Beilstein: 4-26.3109 · Syst.-Nr. 4026
CAS-Reg.No.: 262-76-0
IUPAC: TP-5.3

–, octadehydro- = Corrole
–, 1,19-seco- = Bilin, 1,2,3,7,8,12,-13,17,18,19-decahydro-

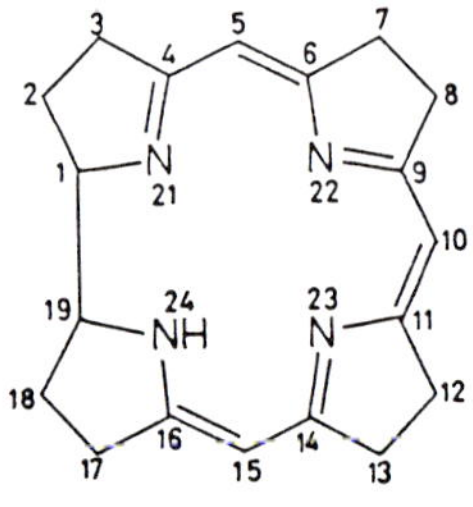

Corrole **Corrol**

sys[CA]: Corrin, octadehydro-

$C_{19}H_{14}N_4$
Beilstein · Syst.-Nr. 4030
CAS-Reg.No.: 26444-09-7
IUPAC: TP-5.3

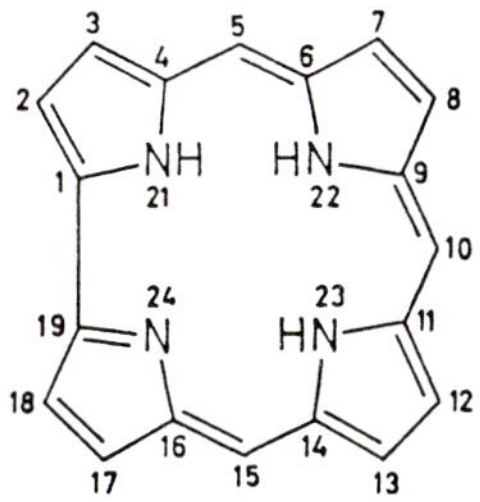

Corticosterone **Corticosteron**

= Pregn-4-ene-3,20-dione,
11β,21-dihydroxy-

Cortisol **Cortisol**

= Pregn-4-ene-3,20-dione,
11β,17,21-trihydroxy-

Cortisone **Cortison**

= Pregn-4-ene-3,11,20-trione,
17,21-dihydroxy-

Corynan **Corynan**

$C_{19}H_{26}N_2$
Beilstein: 4-22.4340* · Syst.-Nr. 3485
CAS-Reg.No.: 1669-07-4
IUPAC: F-2

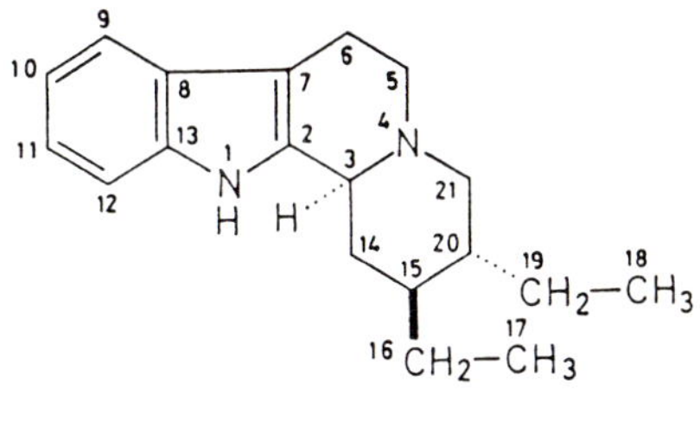

Corynoxan **Corynoxan**

$C_{19}H_{28}N_2$
Beilstein: 4-23.2653* · Syst.-Nr. 3484
CAS-Reg.No.: 34159-89-2
IUPAC: F-2

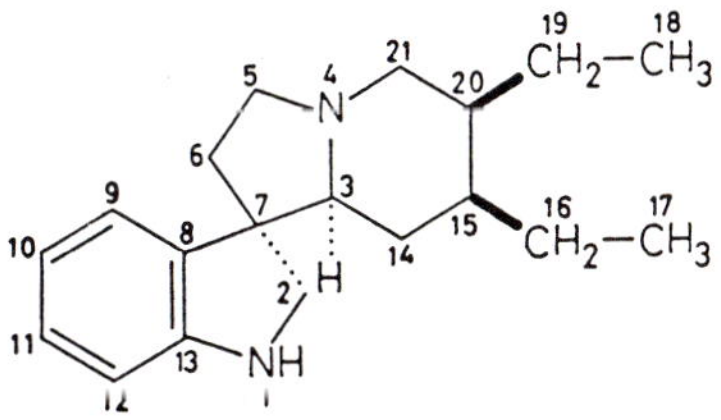

Coumarin **Cumarin**

sys[B]: Chromen-2-on
[CA]: 2*H*-1-Benzopyran-2-one

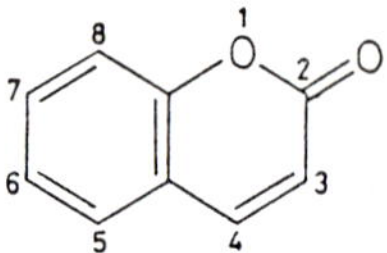

$C_9H_6O_2$
Beilstein: 4-17.5055 · Syst.-Nr. 2464
CAS-Reg.No.: 91-64-5
IUPAC: C-473.1

Cresol **Kresol**

sys[CA]: Phenol, methyl-

C_7H_8O
CAS-Reg.No.: 1319-77-3
IUPAC: C-202.2

m-

o-Cresol
Beilstein: 4-6.1940 · Syst.-Nr. 525
CAS-Reg.No.: 95-48-7

m-Cresol
Beilstein: 4-6.2035 · Syst.-Nr. 526
CAS-Reg.No.: 108-39-4

p-Cresol
Beilstein: 4-6.2093 · Syst.-Nr. 527
CAS-Reg.No.: 106-44-5

Crinan **Crinan**

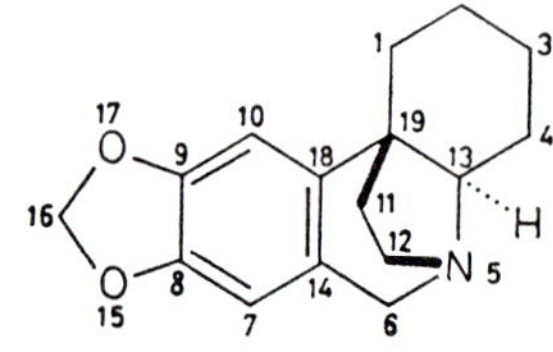

$C_{16}H_{19}NO_2$
Beilstein: 4-27.6332 · Syst.-Nr. 4404
CAS-Reg.No.: 510-70-3
IUPAC: F-2

Crotalanan **Crotalanan**

$C_{17}H_{27}NO_2$
Beilstein: 4-21.250* · Syst.-Nr. 4401
CAS-Reg.No.: 39608-68-9
IUPAC: F-2

Crotonic acid **Crotonsäure**

sys[CA]: 2-Butenoic acid

$C_4H_6O_2$
Beilstein · Syst.-Nr. 163
CAS-Reg.No.: 3724-65-0
IUPAC: C-404.1

$CH_3-CH=CH-CO-OH$

–, 3-methyl- = *Senecioic acid*

cis-Crotonic acid
Beilstein: 4-2.1497
CAS-Reg.No.: 503-64-0
–, 2-methyl- = *Angelic acid*

trans-Crotonic acid
Beilstein: 4-2.1498
CAS-Reg.No.: 107-93-7
–, 2-methyl- = *Tiglic acid*

Cubane **Cuban**

sys[CA]: Pentacyclo[4.2.0.0^{2,5}.0^{3,8}.0^{4,7}]-octane

C_8H_8
Beilstein · Syst.-Nr. 473
CAS-Reg.No.: 277-10-1

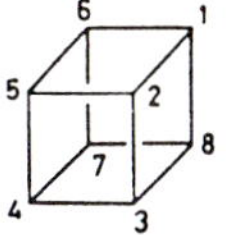

Cucurbitane **Cucurbitan**

sys[CA]: 19-Norlanostane, 9-methyl-, (5α,9β,10α)-

$C_{30}H_{54}$
Beilstein: 3-8.4377* · Syst.-Nr. 472
CAS-Reg.No.: 65441-59-0
IUPAC: 2S-2.3(31)

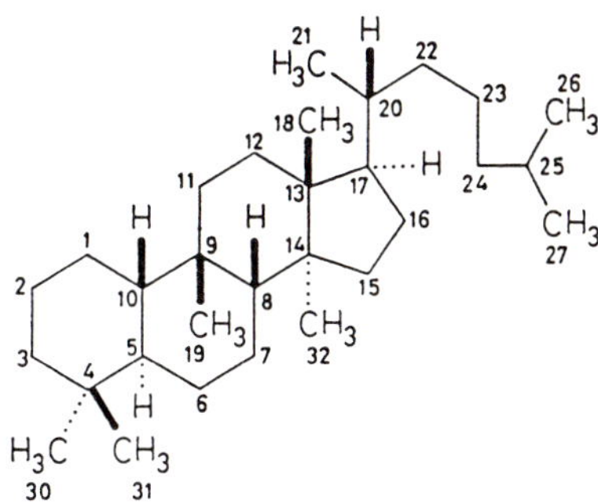

Cumene **Cumol**

sys[CA]: Benzene, (1-methylethyl)-

C_9H_{12}
Beilstein: 4-5.985 · Syst.-Nr. 468
CAS-Reg.No.: 98-82-8
IUPAC: A-12.1

Curan **Curan**

$C_{19}H_{26}N_2$
Beilstein: 4-23.1428* · Syst.-Nr. 3485
CAS-Reg.No.: 34174-79-3
IUPAC: F-2

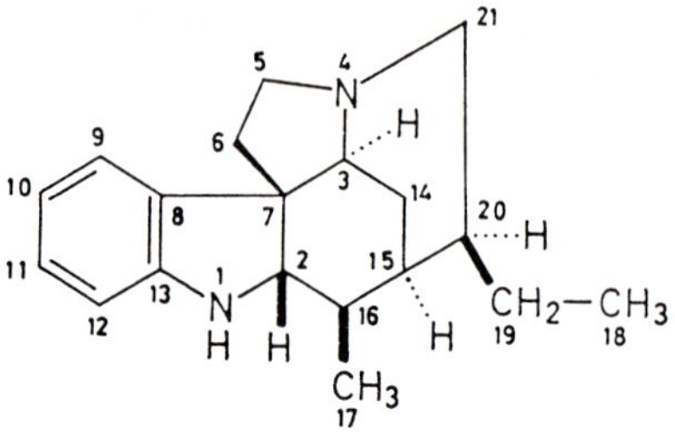

Cycleanan **Cycleanan**

$C_{32}H_{30}N_2O_2$
Beilstein · Syst.-Nr. 4635
IUPAC: F-2

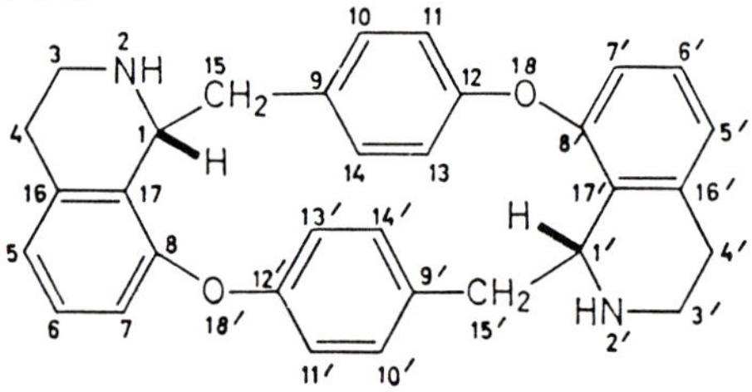

Cycleanine **Cycleanin**

= Cycleanan, 6,7,6′,7′-tetramethoxy-2,2′-dimethyl-

Cycloartane **Cycloartan**

= Lanostane, (5α)-9β,19-cyclo-

Cymarose **Cymarose**

= *ribo*-2,6-Dideoxy-hexose, O^3-methyl-

Cymene **Cymol**

sys[CA]: Benzene, methyl-(1-methyl-ethyl)-

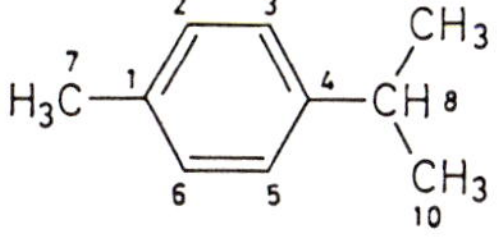

p-

$C_{10}H_{14}$
Beilstein · Syst.-Nr. 469
CAS-Reg.No.: 25155-15-1
IUPAC: A-12.1

o-Cymene
Beilstein: 4-5.1057
CAS-Reg.No.: 527-84-4

m-Cymene
Beilstein: 4-5.1058
CAS-Reg.No.: 535-77-3

p-Cymene
Beilstein: 4-5.1060
CAS-Reg.No.: 99-87-6

Cystathionine **Cystathionin**

sys[B]: Homocystein, *S*-[2-Amino-2-carboxy-äthyl]-
[CA]: Homocysteine, *S*-(2-amino-2-carboxyethyl)-

$C_7H_{14}N_2O_4S$

$$HO-CO-CH(NH_2)-CH_2-CH_2-S-CH_2-CH(NH_2)-CO-OH$$

Beilstein · Syst.-Nr. 376
CAS-Reg.No.: 6899-07-6
IUPAC: C-421.1; 3AA-Appendix

D,D-Cystathionine
Beilstein: 4-4.3197
CAS-Reg.No.: 31982-11-3

D,L-Cystathionine
Beilstein: 4-4.3197
CAS-Reg.No.: 2867-15-4

L,L-Cystathionine
Beilstein: 4-4.3197
CAS-Reg.No.: 56-88-2

L,D-Cystathionine
Beilstein: 4-4.3198

DL,DL-Cystathionine
Beilstein: 4-4.3197
CAS-Reg.No.: 535-34-2

Cysteic acid **Cysteinsäure**

= Propionic acid, 2-amino-3-sulfo-

Cysteine **Cystein**

$C_3H_7NO_2S$
Beilstein · Syst.-Nr. 376
CAS-Reg.No.: 4371-52-2
IUPAC: 3AA-1

$HS-CH_2-CH(NH_2)-CO-OH$

D-Cysteine
Beilstein: 3-4.1618
CAS-Reg.No.: 921-01-7
L-Cysteine
Beilstein: 4-4.3144
CAS-Reg.No.: 52-90-4
DL-Cysteine
Beilstein: 4-4.3145
CAS-Reg.No.: 3374-22-9

Cystine **Cystin**
sys[B]: Propionsäure, 2,2′-Diamino-
3,3′-disulfandiyl-di-

$C_6H_{12}N_2O_4S_2$

$$HO-CO-CH(NH_2)-CH_2-S-S-CH_2-CH(NH_2)-CO-OH$$

Beilstein · Syst.-Nr. 376
CAS-Reg.No.: 24645-67-8
IUPAC: C-421.1; 3AA-Appendix

D-Cystine
Beilstein: 4-4.3155
CAS-Reg.No.: 349-46-2
L-Cystine
Beilstein: 4-4.3155
CAS-Reg.No.: 56-89-3
DL-Cystine
Beilstein: 4-4.3156
CAS-Reg.No.: 923-32-0
meso-Cystine
Beilstein: 3-4.1621
CAS-Reg.No.: 6020-39-9

Cytidine **Cytidin**

$C_9H_{13}N_3O_5$
Beilstein: 4-25.3667 · Syst.-Nr. 3774
CAS-Reg.No.: 65-46-3

Cytisine **Cytisin**

sys[CA]: 1,5-Methano-8*H*-pyrido[1,2-*a*]-[1,5]diazocin-8-one, 1,2,3,4,5,6-hexahydro-, (1*R*)-

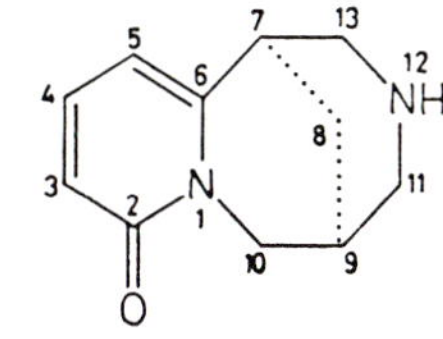

$C_{11}H_{14}N_2O$
Beilstein: 4-24.321 · Syst.-Nr. 3567
CAS-Reg.No.: 485-35-8

Cytochalasan **Cytochalasan**

Cytochalasans are (3*S*)-3*r*,4*t*,5*t*-Trimethyl-(3a*c*,6a*c*)-perhydro-cycloalk-[*d*]isoindoles; the sizes of additional cycloalkane rings, fused to the basic skeleton, are given in square brackets before the cyclochalasan name.

[11]Cytochalasan

$C_{20}H_{37}N$
Beilstein · Syst.-Nr. 3048
CAS-Reg.No.: 51041-85-1
IUPAC: F-2

[13]Cytochalasan

$C_{22}H_{41}N$
Beilstein · Syst.-Nr. 3048
CAS-Reg.No.: 51041-84-0
IUPAC: F-2

Cytoporphyrin **Cytoporphyrin**

syn: Porphyrin a

sys[CA]: 21*H*,23*H*-Porphine-2,18-dipropanoic acid, 7-ethenyl-17-formyl-12-(1-hydroxy-5,9,13-trimethyl-4,8,12-tetradecatrienyl)-3,8,13-trimethyl-, (*E*,*E*)-

$C_{49}H_{58}N_4O_6$
Beilstein: 4-26.3338 · Syst.-Nr. 4173
CAS-Reg.No.: 25162-02-1
IUPAC: TP-2.1

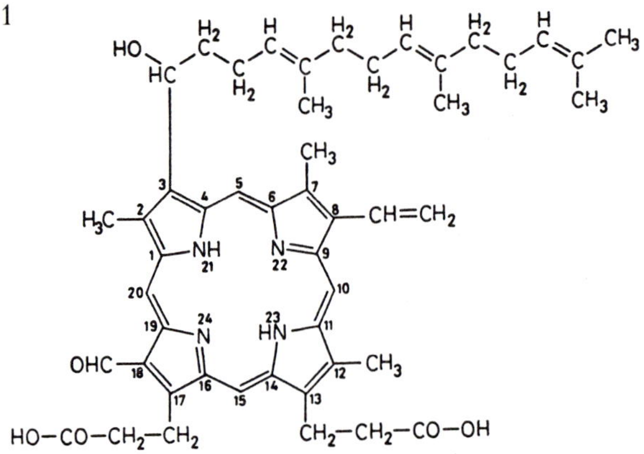

Cytosine **Cytosin**

sys[CA]: 2(1*H*)-Pyrimidinone, 4-amino-

$C_4H_5N_3O$
Beilstein: 4-25.3654 · Syst.-Nr. 3774
CAS-Reg.No.: 71-30-7

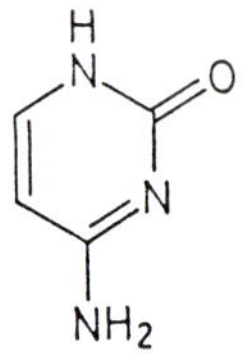

Dammarane **Dammaran**

$C_{30}H_{54}$
Beilstein: 3-6.2717* · Syst.-Nr. 472
CAS-Reg.No.: 545-22-2
IUPAC: 2S-2.3(30)

–, (8α,9β,13α,14β)- = *Protostane*

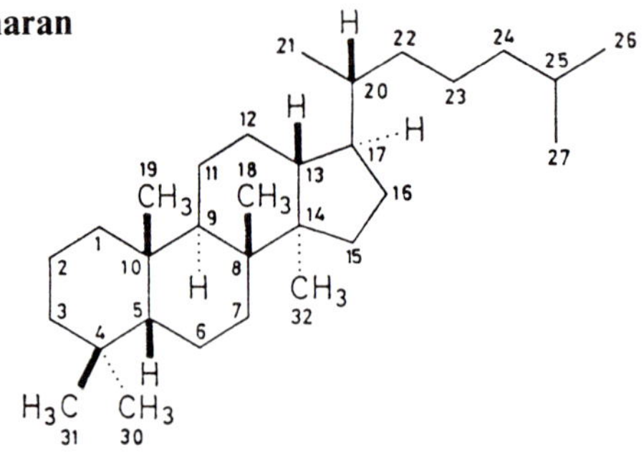

Daphnane **Daphnan**

$C_{22}H_{37}N$
Beilstein · Syst.-Nr. 3065
CAS-Reg.No.: 39686-16-3
IUPAC: F-2

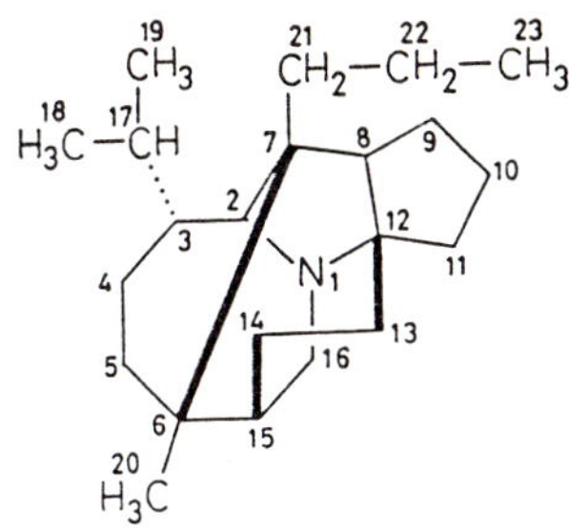

Dasycarpidan **Dasycarpidan**

$C_{17}H_{22}N_2$
Beilstein: 4-23.1423* · Syst.-Nr. 3485
CAS-Reg.No.: 41180-88-5
IUPAC: F-2

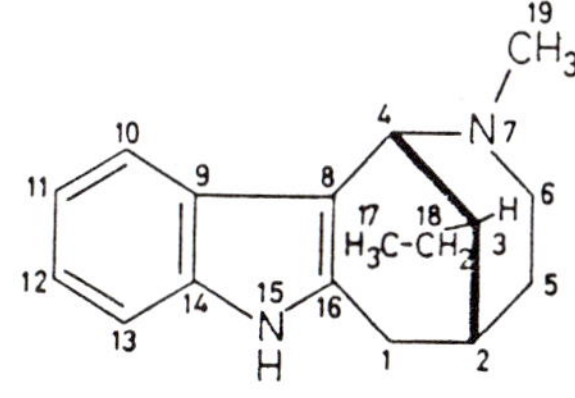

Dendrobane **Dendroban**

$C_{16}H_{25}NO$
Beilstein: 4-22.2132* · Syst.-Nr. 4194
CAS-Reg.No.: 39608-76-9
IUPAC: F-2

Deoxybenzoin **Desoxybenzoin**
sys[CA]: Ethanone, 1,2-diphenyl-

$C_{14}H_{12}O$
Beilstein: 4-7.1393 · Syst.-Nr. 653
CAS-Reg.No.: 451-40-1
IUPAC: C-313.2

Deoxyretinol **Desoxyretinol**

syn: Axerophthene
sys[B]: Nona-1,3,5,7-tetraen, 3,7-Dimethyl-1-[2,6,6-trimethyl-cyclohex-1-enyl]-
[CA]: Cyclohexene, 2-(3,7-dimethyl-1,3,5,7-nonatetraenyl)-1,3,3-trimethyl-

$C_{20}H_{30}$
Beilstein: 4-5.1611 · Syst.-Nr. 474
CAS-Reg.No.: 564-86-3
IUPAC: Ret-4.1

Deuteroporphyrin **Deuteroporphyrin**

syn: Deuteroporphyrin IX
sys[CA]: 21*H*,23*H*-Porphine-2,18-dipropanoic acid, 3,7,12,17-tetramethyl-

$C_{30}H_{30}N_4O_4$
Beilstein: 4-26.2992 · Syst.-Nr. 4173
CAS-Reg.No.: 448-65-7
IUPAC: TP-2.1

Diginane **Diginan**

sys[CA]: Pregnane, (5α,14β,17α)-

$C_{21}H_{36}$
Beilstein: 3-5.1120 · Syst.-Nr. 472
CAS-Reg.No.: 78964-23-5

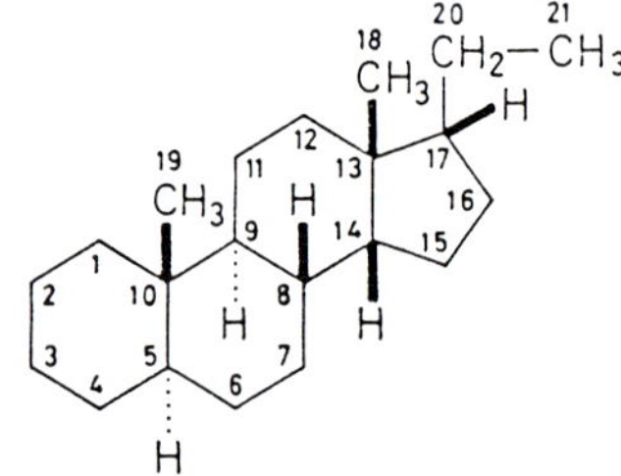

Diginose **Diginose**

= *lyxo*-2,6-Dideoxy-hexose, O^3-methyl-

Digitalose **Digitalose**

= Galactose, O^3-methyl-6-deoxy-

Digitoxigenin **Digitoxigenin**

= Card-20(22)-enolide,
3β,14-dihydroxy-(5β,14β)-

Digitoxose **Digitoxose**

= D-*ribo*-2,6-Dideoxy-hexose

Dithiobiuret **Dithiobiuret**

sys[CA]: Thioimidodicarbonic diamide

$C_2H_5N_3S_2$
Beilstein: 4-3.356 · Syst.-Nr. 216
CAS-Reg.No.: 541-53-7
IUPAC: C-975.1

$H_2N-CS-NH-CS-NH_2$

Dopa **Dopa**

= L-Tyrosine, 3-hydroxy-

Drimane **Driman**

sys[CA]: Naphthalene, decahydro-
1,1,4a,5,6-pentamethyl-,
[4a*R*-(4aα,5α,6α,8aβ)]-

$C_{15}H_{28}$
Beilstein: 3-9.273* · Syst.-Nr. 453
CAS-Reg.No.: 5951-58-6
IUPAC: F-2

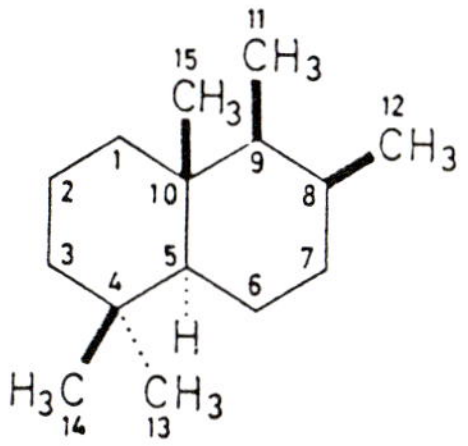

Eburicane **Eburican**

syn: Laudane
sys[CA]: Lanostane, 24-methyl-, (24*S*)-

$C_{31}H_{56}$
Beilstein: 4-5.1242 · Syst.-Nr. 472

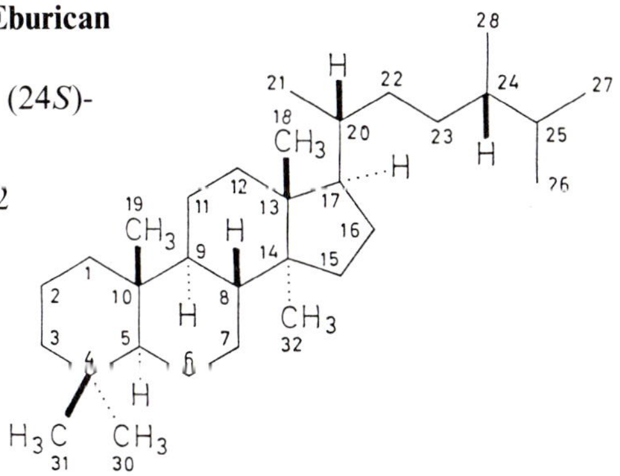

Eburnamenine **Eburnamenin**

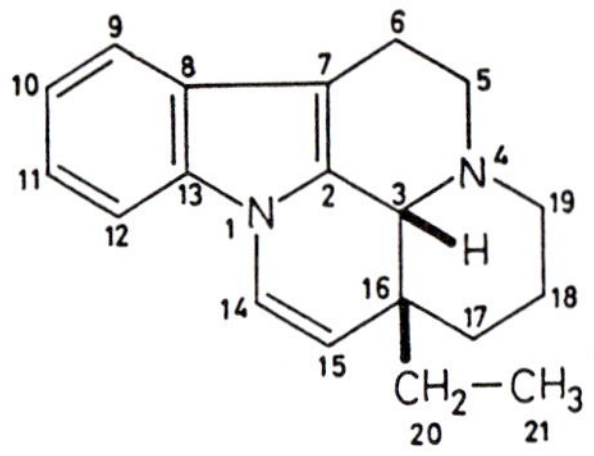

$C_{19}H_{22}N_2$
Beilstein: 4-23.1789 · Syst.-Nr. 3487
CAS-Reg.No.: 517-30-6

Elaidic acid **Elaidinsäure**

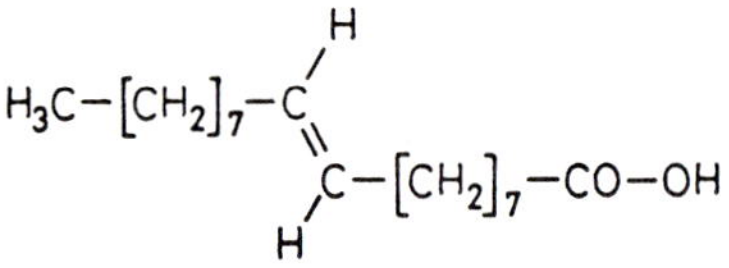

sys[B]: Octadec-9*t*-ensäure
[CA]: 9-Octadecenoic acid, (*E*)-

$C_{18}H_{34}O_2$
Beilstein: 4-2.1647 · Syst.-Nr. 178
CAS-Reg.No.: 112-79-8
IUPAC: C-404.1

Eleocarpine **Eleocarpin**

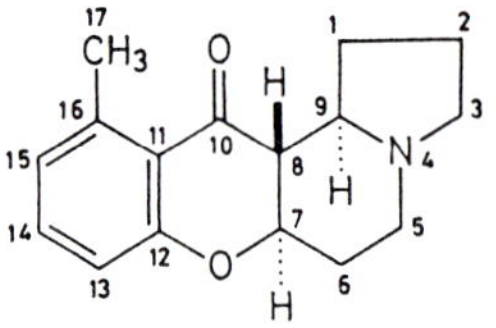

$C_{16}H_{19}NO_2$
Beilstein · Syst.-Nr. 4280
CAS-Reg.No.: 30891-90-8

Emetan **Emetan**

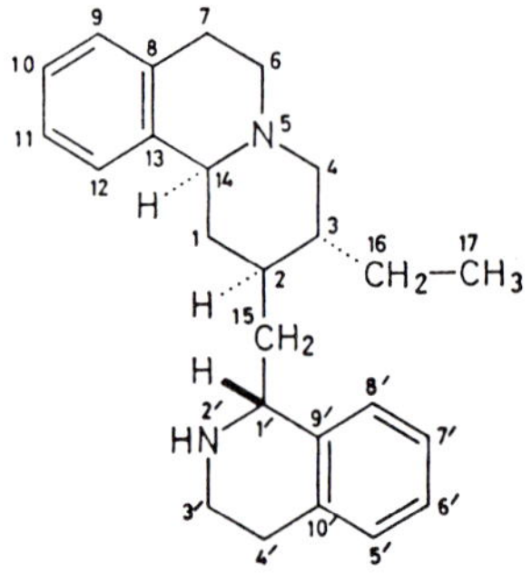

$C_{25}H_{32}N_2$
Beilstein: 4-22.6148* · Syst.-Nr. 3488
CAS-Reg.No.: 36506-59-9

Epicellobiose **Epicellobiose**

= D-Mannose, O^4-β-D-glucopyranosyl-

Epigentiobiose **Epigentiobiose**

= D-Mannose, O^6-β-D-glucopyranosyl-

Epilactose **Epilactose**

= D-Mannose, O^4-β-D-galactopyranosyl-

Epimelibiose **Epimelibiose**

= D-Mannose, O^6-α-D-galactopyranosyl-

Ercalciol **Ercalciol**

syn: Ergocalciferol, Vitamin D_2

sys[CA]: 9,10-Secoergosta-5,7,10(19),-22-tetraen-3-ol, (3β,5*Z*,7*E*,-22*E*)-

$C_{28}H_{44}O$

Beilstein: 4-6.4404 · Syst.-Nr. 538 A

CAS-Reg.No.: 50-14-6

IUPAC: VD-6

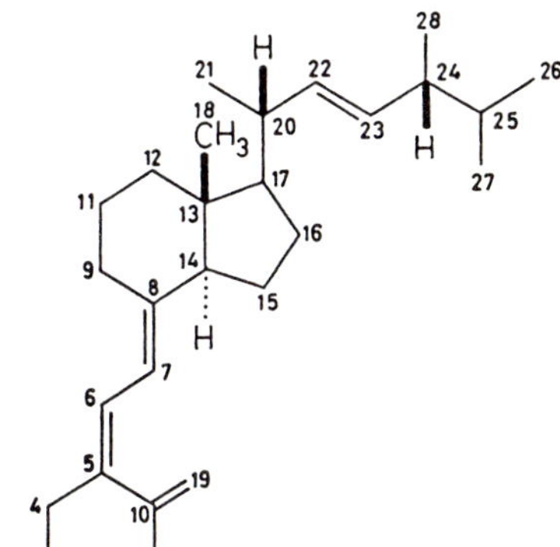

Eremophilane **Eremophilan**

sys[CA]: Naphthalene, decahydro-1,8a-dimethyl-7-(1-methylethyl)-, [1*S*-(1α,4aα,7α,8aα)]-

$C_{15}H_{28}$

Beilstein: 4-5.355 · Syst.-Nr. 453

CAS-Reg.No.: 3242-05-5

IUPAC: F-(33)

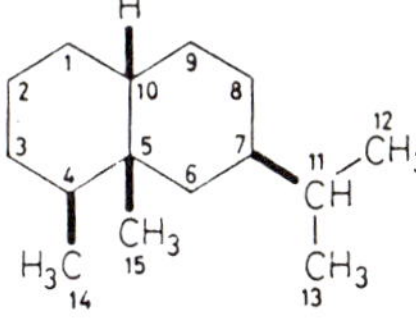

Ergocalciferol **Ergocalciferol**

= Ercalciol

Ergoline **Ergolin**

syn: Ergoline I

$C_{14}H_{16}N_2$

Beilstein: 4-22.6496* · Syst.-Nr. 3485

CAS-Reg.No.: 478-88-6

Ergoline I **Ergolin-I**

= Ergoline

Ergostane **Ergostan**

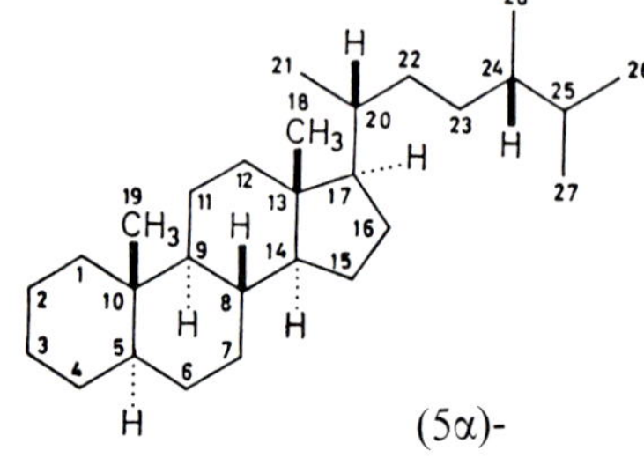

$C_{28}H_{50}$
Beilstein: 4-5.1234 · Syst.-Nr. 472
CAS-Reg.No.: 25318-39-2
IUPAC: 2S-2.3(20)

–, (5α,9β,10α)- = Lumistane

Ergosterol **Ergosterin**

= Ergosta-5,7,22-trien-3β-ol

Ergotaman **Ergotaman**

$C_{25}H_{33}N_5O$
Beilstein: 4-25.907* · Syst.-Nr. 4607
CAS-Reg.No.: 34612-42-5

Erythrinan **Erythrinan**

$C_{16}H_{21}N$
Beilstein: 4-20.2878* · Syst.-Nr. 3082
CAS-Reg.No.: 27711-98-4
IUPAC: F-2

Erythrose **Erythrose**

sys[CA]: Butanal, 2,3,4-trihydroxy-, (*R**,*R**)-

$C_4H_8O_4$
Beilstein · Syst.-Nr. 124
CAS-Reg.No.: 1758-51-6
IUPAC: Carb-5

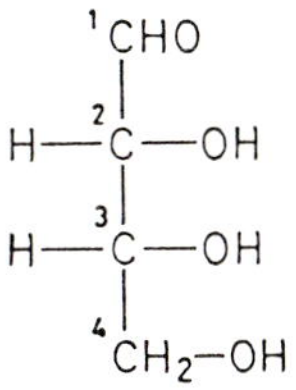

D-Erythrose
Beilstein: 4-1.4172
CAS-Reg.No.: 583-50-6
L-Erythrose
Beilstein: 4-1.4172
CAS-Reg.No.: 533-49-3
DL-Erythrose
Beilstein: 4-1.4173
CAS-Reg.No.: 29825-68-1

D-

Essigsäure = Acetic acid

Estrane **Östran**

$C_{18}H_{30}$
Beilstein: 3-5.1108 · Syst.-Nr. 472
CAS-Reg.No.: 24749-37-9
IUPAC: 2S-2.2(18)

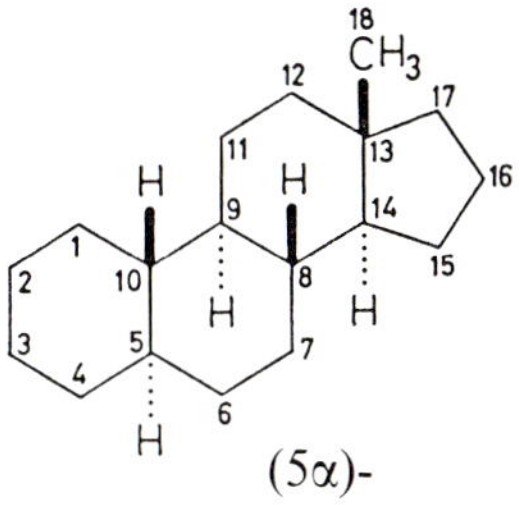

(5α)-

Estriol **Östriol**

= Estra-1,3,5(10)-triene-3,16α,17β-triol

Estrone **Östron**

= Estra-1,3,5(10)-trien-17-one, 3-hydroxy-

Ethane **Äthan**

C_2H_6
Beilstein: 4-1.108 · Syst.-Nr. 7
CAS-Reg.No.: 74-84-0
IUPAC: A-1.1

H_3C-CH_3

Ethylene glycol **Äthylenglykol**

sys[CA]: 1,2-Ethanediol

$C_2H_6O_2$
Beilstein: 4-1.2369 · Syst.-Nr. 30
CAS-Reg.No.: 107-21-1
IUPAC: C-201.4

$HO-CH_2-CH_2-OH$

Etioporphyrin I **Ätioporphyrin-I**

sys[CA]: 21*H*,23*H*-Porphine, 2,7,12,17-tetraethyl-3,8,13,18-tetramethyl-

$C_{32}H_{38}N_4$
Beilstein: 4-26.1915 · Syst.-Nr. 4031
CAS-Reg.No.: 448-71-5
IUPAC: TP-2.1/2.2

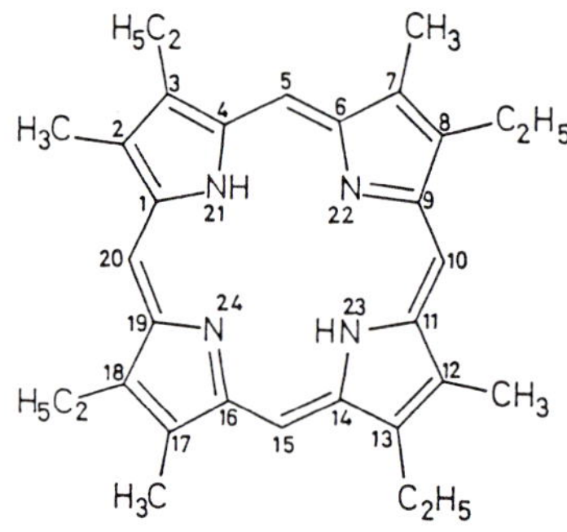

Eudesmane **Eudesman**

sys[CA]: Naphthalene, decahydro-1,4a-dimethyl-7-(1-methylethyl)-, [1*R*-(1α,4aβ,7β,8aα)]-

$C_{15}H_{28}$
Beilstein: 4-5.355 · Syst.-Nr. 453
CAS-Reg.No.: 473-11-0

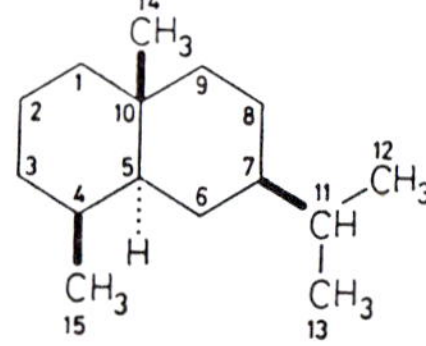

Eugenol **Eugenol**

sys[B]: Phenol, 4-Allyl-2-methoxy-
[CA]: Phenol, 2-methoxy-4-(2-propenyl)-

$C_{10}H_{12}O_2$
Beilstein: 4-6.6337 · Syst.-Nr. 560
CAS-Reg.No.: 97-53-0
IUPAC: C-214.1

Euphane **Euphan**

sys[CA]: Lanostane, (13α,14β,17α)-

$C_{30}H_{54}$
Beilstein: 3-5.1338 · Syst.-Nr. 472
CAS-Reg.No.: 516-29-0
IUPAC: 2S-2.3(29)

Euphorbane **Euphorban**

sys[CA]: Lanostane, 24-methyl-, [20*S*,24*R*-(13α,14β,17α)]-

$C_{31}H_{56}$
Beilstein: 3-6.2733* · Syst.-Nr. 472

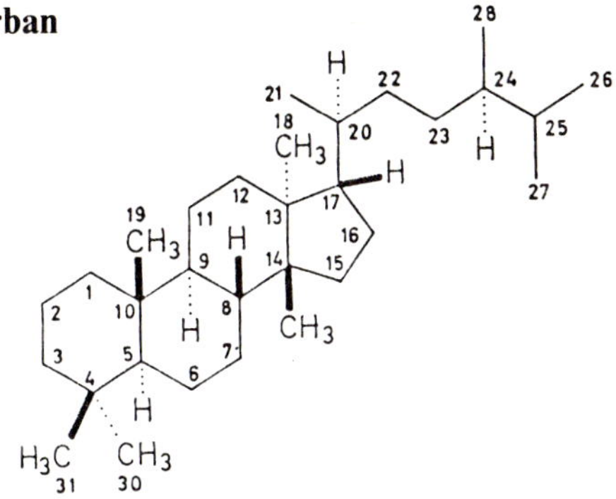

Evonimine **Evonimin**

$C_{36}H_{43}NO_{17}$
Beilstein: 4-27.6847 · Syst.-Nr. 4475
CAS-Reg.No.: 41758-54-7

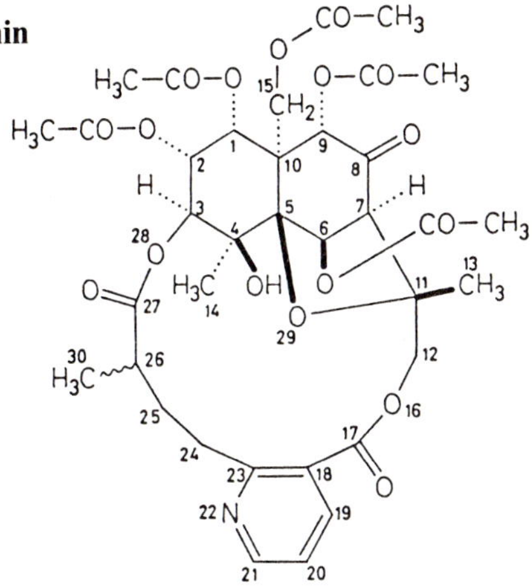

Evonine **Evonin**

$C_{36}H_{43}NO_{17}$
Beilstein: 4-27.6848 · Syst.-Nr. 4475
CAS-Reg.No.: 33458-64-9

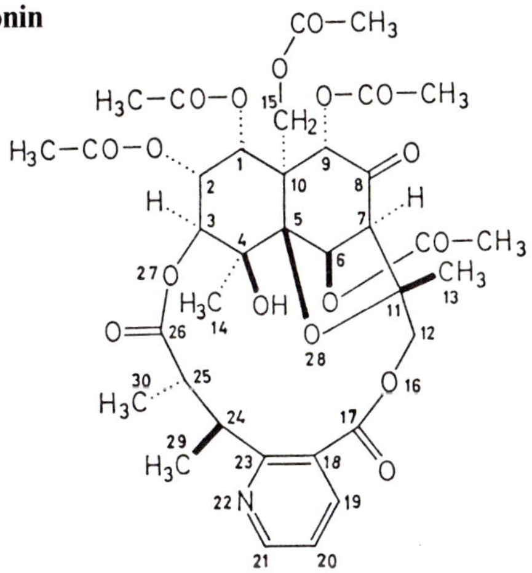

Fawcettidine **Fawcettidin**

$C_{16}H_{23}NO$
Beilstein · Syst.-Nr. 3183
CAS-Reg.No.: 14912-31-3

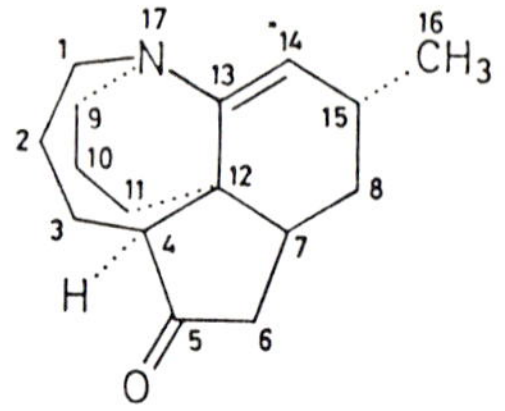

Fernane **Fernan**

= Gammacerane, *D:C*-friedo-*B′:A′*-neo-

Fesane **Fesan**

sys[CA]: 16,23-Cyclocholestane

$C_{27}H_{46}$
Beilstein: 3-6.6477* · Syst.-Nr. 473
CAS-Reg.No.: 38776-69-1

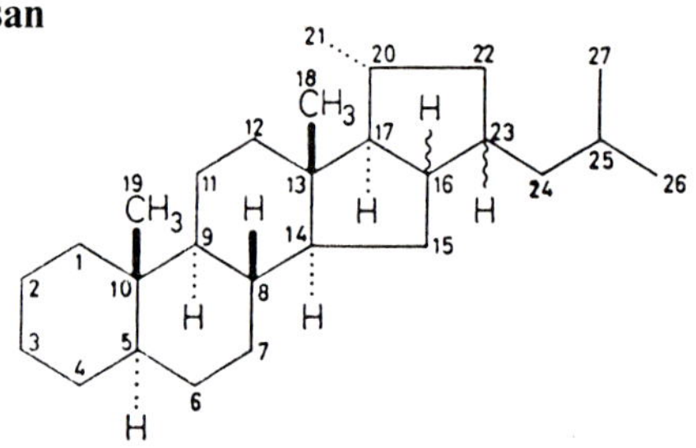

Flavan **Flavan**

= Chroman, 2-phenyl-

Flavene **Flaven**

= Chromene, 2-phenyl-

Fluoranthene **Fluoranthen**

$C_{16}H_{10}$
Beilstein: 4-5.2463 · Syst.-Nr. 487
CAS-Reg.No.: 206-44-0
IUPAC: A-21.2(14)

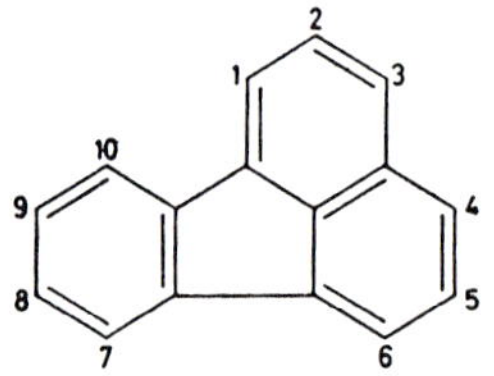

Fluorene **Fluoren**

sys[CA]: 9*H*-Fluorene

$C_{13}H_{10}$
Beilstein: 4-5.2142 · Syst.-Nr. 480
CAS-Reg.No.: 86-73-7
IUPAC: A-21.2(10)

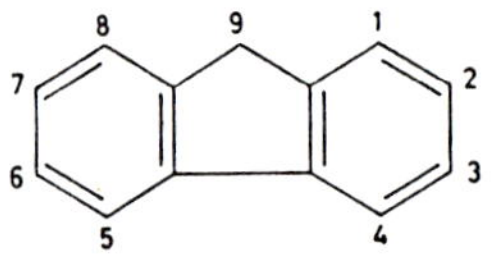

Formaldehyde **Formaldehyd**

CH_2O
Beilstein: 4-1.3017 · Syst.-Nr. 74
CAS-Reg.No.: 50-00-0
IUPAC: C-305.1

H – CHO

Formic acid **Ameisensäure**

CH_2O_2
Beilstein: 4-2.3 · Syst.-Nr. 155
CAS-Reg.No.: 64-18-6
IUPAC: C-404.1

H – CO – OH

Formosanan **Formosanan**

$C_{18}H_{22}N_2O$
Beilstein: 4-27.7359* · Syst.-Nr. 4494
CAS-Reg.No.: 34723-40-5

Friedelane **Friedelan**

sys[CA]: *D:A*-Friedooleanane

$C_{30}H_{52}$
Beilstein: 3-5.1341 · Syst.-Nr. 473
CAS-Reg.No.: 559-73-9

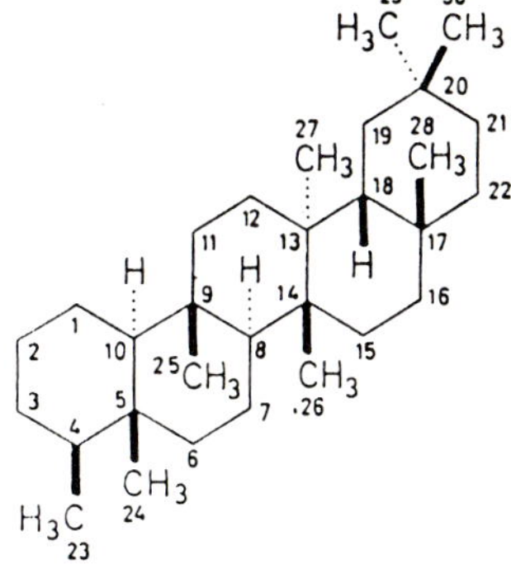

Fructofuranose **Fructofuranose**

$C_6H_{12}O_6$
Beilstein · Syst.-Nr. 145
IUPAC: Carb-10/18

α-D-Fructofuranose
Beilstein: 4-1.4402
CAS-Reg.No.: 10489-79-9

β-D-Fructofuranose
Beilstein: 4-1.4402
CAS-Reg.No.: 470-23-5

β-D-

Fructopyranose **Fructopyranose**

$C_6H_{12}O_6$
Beilstein · Syst.-Nr. 145
IUPAC: Carb-10/18

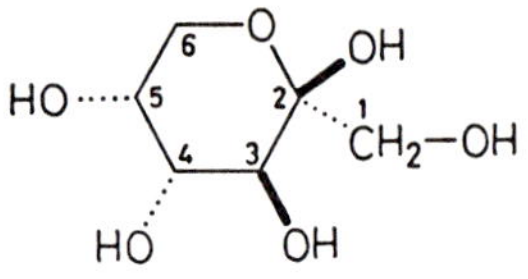

α-D-Fructopyranose
Beilstein: 4-1.4402
CAS-Reg.No.: 10489-81-3
β-D-Fructopyranose
Beilstein: 4-1.4402
CAS-Reg.No.: 7660-25-5
β-L-Fructopyranose
CAS-Reg.No.: 41612-84-4

β-D-

Fructose **Fructose**

$C_6H_{12}O_6$
Beilstein · Syst.-Nr. 145
CAS-Reg.No.: 30237-26-4
IUPAC: Carb-10

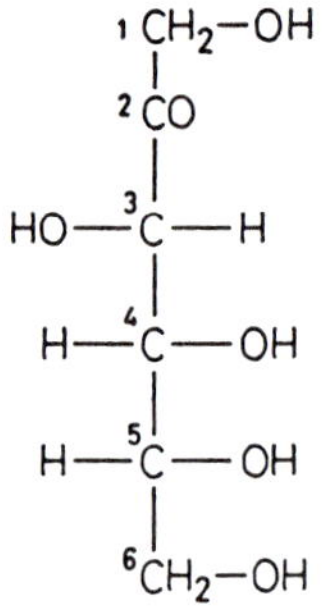

D-Fructose
Beilstein: 4-1.4401
CAS-Reg.No.: 57-48-7
–, O^3-α-D-glucopyranosyl- = *Turanose*

L-Fructose
Beilstein: 4-1.4411
CAS-Reg.No.: 7776-48-9
DL-Fructose
Beilstein: 4-1.4411
CAS-Reg.No.: 6035-50-3

D-

Fucose **Fucose**

= Galactose, 6-deoxy-

Fulvene **Fulven**

sys[CA]: 1,3-Cyclopentadiene, 5-methylene-

C_6H_6
Beilstein: 4-5.764 · Syst.-Nr. 465 A
CAS-Reg.No.: 497-20-1
IUPAC: A-61.6

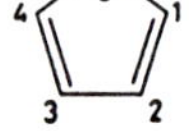

Fumaric acid **Fumarsäure**

sys[CA]: 2-Butenedioic acid, (*E*)-

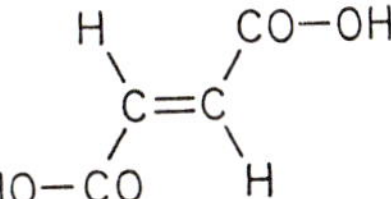

$C_4H_4O_4$
Beilstein: 4-2.2202 · Syst.-Nr. 179
CAS-Reg.No.: 110-17-8
IUPAC: C-404.1

–, methyl- = *Mesaconic acid*

Furan **Furan**

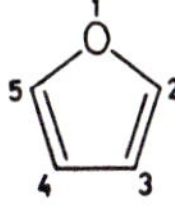

C_4H_4O
Beilstein: 5-17/1.291 · Syst.-Nr. 2364
CAS-Reg.No.: 110-00-9
IUPAC: B-2.11(5)

Furazan **Furazan**

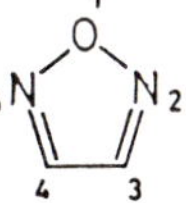

$C_2H_2N_2O$
Beilstein · Syst.-Nr. 4488
CAS-Reg.No.: 288-37-9
IUPAC: B-2.11(46)

Furostan **Furostan**

$C_{27}H_{46}O$
Beilstein: 4-17.470 · Syst.-Nr. 2366
CAS-Reg.No.: 34783-87-4

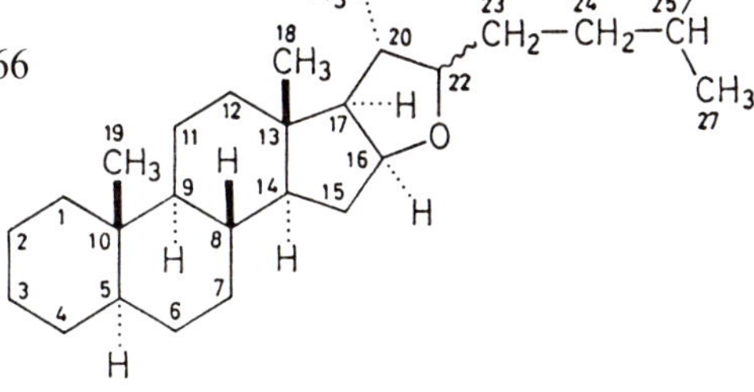

Galactofuranose **Galactofuranose**

$C_6H_{12}O_6$
Beilstein · Syst.-Nr. 144
IUPAC: Carb-5/18

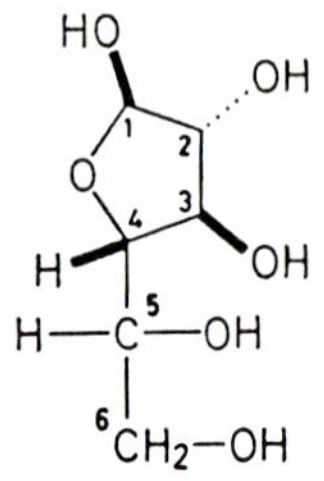

α-D-Galactofuranose
Beilstein: 4-1.4336
CAS-Reg.No.: 36468-82-3
β-D-Galactofuranose
Beilstein: 4-1.4336
CAS-Reg.No.: 7045-51-4
α-L-Galactofuranose
CAS-Reg.No.: 41846-89-3
β-L-Galactofuranose
CAS-Reg.No.: 41846-88-2

β-D-

Galactopyranose **Galactopyranose**

$C_6H_{12}O_6$
Beilstein · Syst.-Nr. 144
CAS-Reg.No.: 40825-89-6
IUPAC: Carb-5/18

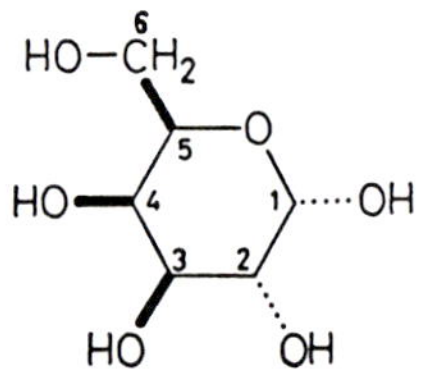

α-D-Galactopyranose
Beilstein: 4-1.4336
CAS-Reg.No.: 3646-73-9
β-D-Galactopyranose
Beilstein: 4-1.4337
CAS-Reg.No.: 7296-64-2
α-L-Galactopyranose
Beilstein: 4-1.4343
CAS-Reg.No.: 12772-65-5
β-L-Galactopyranose
CAS-Reg.No.: 42789-83-3

α-D-

Galactose **Galactose**

$C_6H_{12}O_6$
Beilstein · Syst.-Nr. 144
CAS-Reg.No.: 26566-61-0
IUPAC: Carb-5

–, 6-deoxy- = *Fucose*
–, O^3-methyl-6-deoxy- = *Digitalose*

D-Galactose
Beilstein: 4-1.4336
CAS-Reg.No.: 59-23-4
–, O^6-α-D-galactopyranosyl- = *Swietenose*
–, O^3-β-D-glucopyranosyl- = *Solabiose*
–, O^4-β-D-glucopyranosyl- = *Lycobiose*

D-

L-Galactose
Beilstein: 4-1.4343
CAS-Reg.No.: 15572-79-9
DL-Galactose
Beilstein: 4-1.4344

Galanthaman **Galanthaman**

$C_{15}H_{19}NO$
Beilstein: 4-27.1993* · Syst.-Nr. 4196
IUPAC: F-2

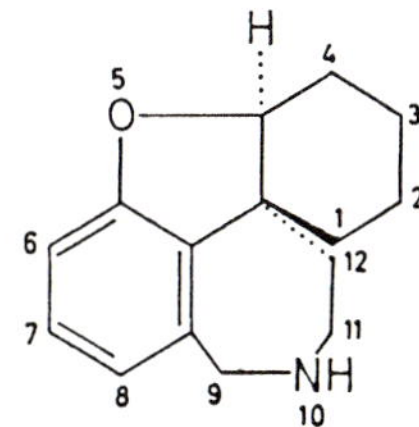

Galanthan **Galanthan**

$C_{15}H_{19}N$
Beilstein: 4-21.2332* · Syst.-Nr. 3082
CAS-Reg.No.: 34159-90-5
IUPAC: F-2

Gallic acid **Gallussäure**
sys[CA]: Benzoic acid, 3,4,5-trihydroxy-

$C_7H_6O_5$
Beilstein: 4-10.1993 · Syst.-Nr. 1136
CAS-Reg.No.: 149-91-7
IUPAC: C-411.1

Gammacerane **Gammaceran**

$C_{30}H_{52}$
Beilstein: 3-6.2731* · Syst.-Nr. 473
CAS-Reg.No.: 559-65-9

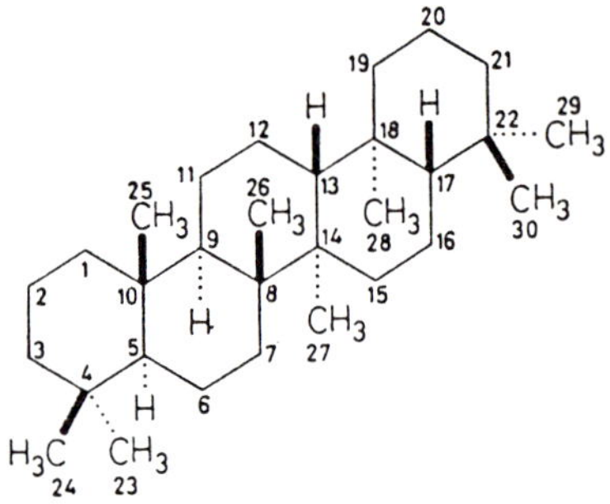

–, 8,14:13,18-diseco- = Ambrane
–, *D:B*-friedo-*B′:A′*-neo- = *Simiarane*
–, *D:C*-friedo-*B′:A′*-neo- = *Fernane*
–, *C*(14a)-homo-27-nor- = *Serratane*
–, *A′*-neo- = *Hopane*
–, 8,14-seco- = Onocerane

Gedunan **Gedunan**

$C_{26}H_{44}O_2$
Beilstein: 4-19.5387* · Syst.-Nr. 2672
IUPAC: F-(42)

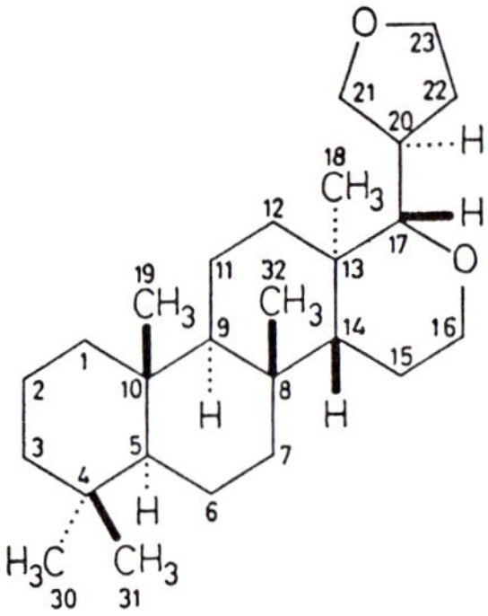

Gelsedine **Gelsedin**

$C_{19}H_{24}N_2O_3$
Beilstein: 4-27.7513 · Syst.-Nr. 4550
CAS-Reg.No.: 7096-96-0

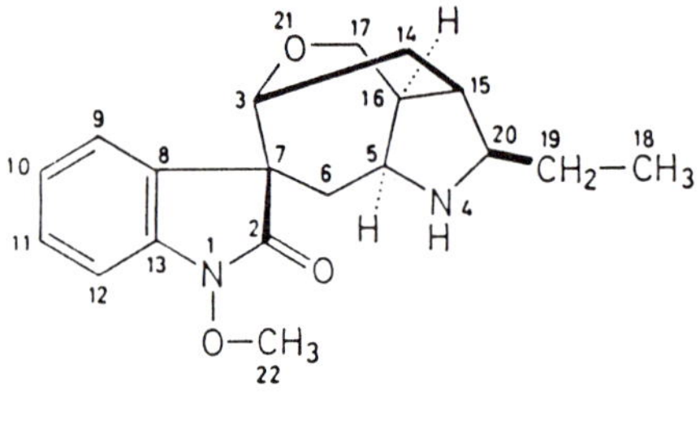

Gelsemine **Gelsemin**

$C_{20}H_{22}N_2O_2$
Beilstein: 4-27.7526 · Syst.-Nr. 4552
CAS-Reg.No.: 509-15-9

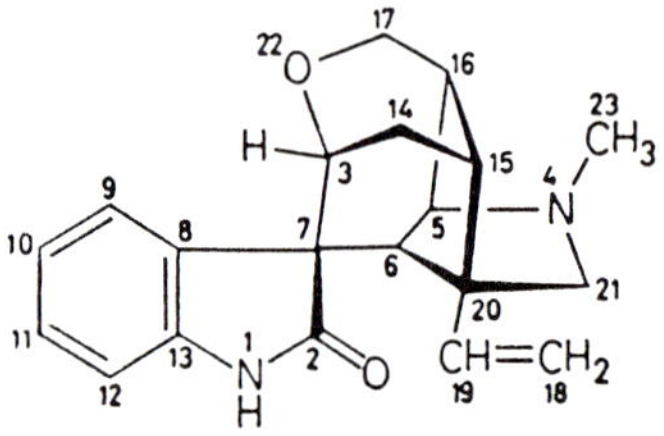

Gentiobiose **Gentiobiose**

= D-Glucose, O^6-β-D-glucopyranosyl-

Geraniol **Geraniol**

sys[CA]: 2,6-Octadien-1-ol, 3,7-dimethyl-, (*E*)-

$C_{10}H_{18}O$
Beilstein: 4-1.2277 · Syst.-Nr. 26
CAS-Reg.No.: 106-24-1
IUPAC: C-201.4

Germacrane **Germacran**

sys[B]: Cyclodecan, (1*S*,7*S*)-4-Isopropyl-1,7-dimethyl-
[CA]: Cyclodecane, 1,7-dimethyl-4-(1-methylethyl)-

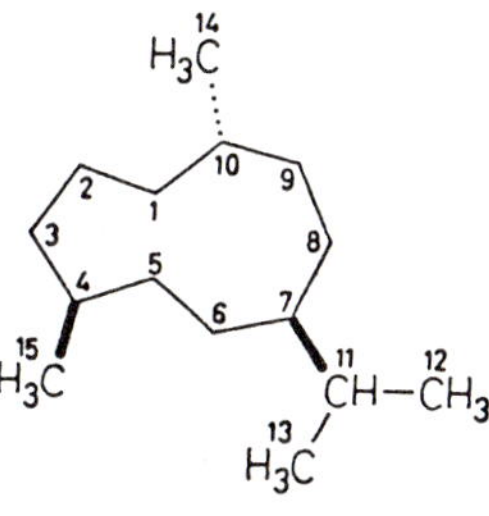

$C_{15}H_{30}$
Beilstein: 4-17.4393* · Syst.-Nr. 452
CAS-Reg.No.: 645-10-3
IUPAC: F-(12)

Gibbane **Gibban**

sys[B]: 7*r*,9a*c*-Methano-benz[*a*]azulen, (7*S*)-(4a*c*,4b*t*,10a*t*)-Dodecahydro-

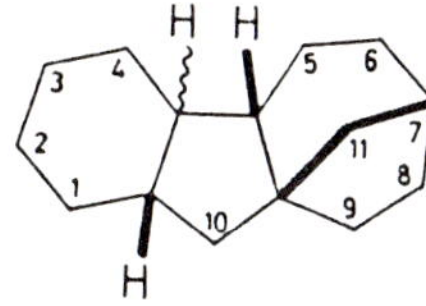

$C_{15}H_{24}$
Beilstein: 4-18.6080* · Syst.-Nr. 471
CAS-Reg.No.: 6902-95-0

Glucofuranose **Glucofuranose**

$C_6H_{12}O_6$
Beilstein · Syst.-Nr. 144
IUPAC: Carb-5/18

α-D-Glucofuranose
CAS-Reg.No.: 36468-84-5
β-D-Glucofuranose
CAS-Reg.No.: 30412-16-9
α-L-Glucofuranose
CAS-Reg.No.: 41846-86-0

β-D-

Glucopyranose **Glucopyranose**

$C_6H_{12}O_6$
Beilstein · Syst.-Nr. 144
IUPAC: Carb-5/18

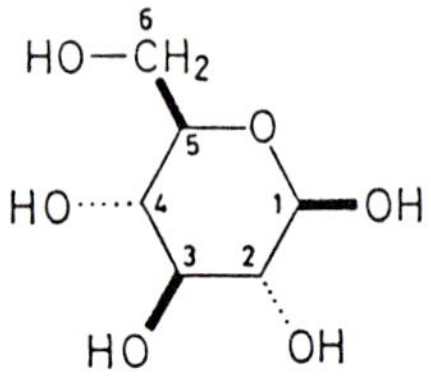

β-D-

α-D-Glucopyranose
Beilstein: 4-1.4304
CAS-Reg.No.: 492-62-6
β-D-Glucopyranose
Beilstein: 4-1.4306
CAS-Reg.No.: 492-61-5
α-L-Glucopyranose
CAS-Reg.No.: 492-66-0
β-L-Glucopyranose
CAS-Reg.No.: 39281-65-7

Glucosamine **Glucosamin**

= D-Glucose, 2-amino-2-deoxy-

Glucose **Glucose**

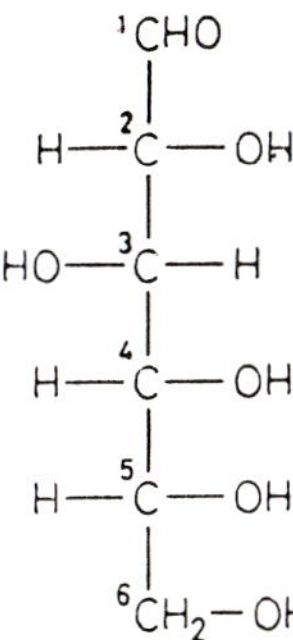

$C_6H_{12}O_6$
Beilstein · Syst.-Nr. 144
IUPAC: Carb-5

–, 6-deoxy- = *Chinovose*
–, O^3-methyl-6-deoxy- = *Thevetose*

D-Glucose
Beilstein: 4-1.4302
CAS-Reg.No.: 50-99-7
–, 2-amino-2-deoxy- = *Glucosamine*
–, O^6-α-L-arabinopyranosyl = *Vicianose*
–, O^6-α-L-(6-deoxy-mannopyranosyl)- = *Rutinose*
–, O^4-β-D-galactopyranosyl- = *Lactose*
–, O^6-α-D-galactopyranosyl- = *Melibiose*
–, O^2-α-D-glucopyranosyl- = *Kojibiose*
–, O^2-β-D-glucopyranosyl- = *Sophorose*
–, O^3-α-D-glucopyranosyl- = *Nigerose*
–, O^3-β-D-glucopyranosyl- = *Laminaribiose*
–, O^4-α-D-glucopyranosyl- = *Maltose*
–, O^4-β-D-glucopyranosyl- = *Cellobiose*
–, O^6-α-D-glucopyranosyl- = *Isomaltose*
–, O^6-β-D-glucopyranosyl- = *Gentiobiose*
–, O^6-α-D-xylopyranosyl- = *Isoprimverose*
–, O^6-β-D-xylopyranosyl- = *Primverose*

L-Glucose
Beilstein: 4-1.4327
CAS-Reg.No.: 921-60-8

DL-Glucose
Beilstein: 4-1.4328
CAS-Reg.No.: 58367-01-4

Glutamic acid **Glutaminsäure**

$C_5H_9NO_4$

$HO-CO-CH_2-CH_2-CH(NH_2)-CO-OH$

Beilstein · Syst.-Nr. 372
CAS-Reg.No.: 6899-05-4
IUPAC: 3AA-1

D-Glutamic acid
Beilstein: 4-4.3028
CAS-Reg.No.: 6893-26-1
L-Glutamic acid
Beilstein: 4-4.3028
CAS-Reg.No.: 56-86-0
DL-Glutamic acid
Beilstein: 4-4.3031
CAS-Reg.No.: 617-65-2

Glutamine **Glutamin**

$C_5H_{10}N_2O_3$

$H_2N-CO-CH_2-CH_2-CH(NH_2)-CO-OH$

Beilstein · Syst.-Nr. 372
CAS-Reg.No.: 6899-04-3
IUPAC: 3AA-1

D-Glutamine
Beilstein: 4-4.3037
CAS-Reg.No.: 5959-95-5
L-Glutamine
Beilstein: 4-4.3038
CAS-Reg.No.: 56-85-9
DL-Glutamine
Beilstein: 4-4.3038
CAS-Reg.No.: 585-21-7

Glutaminsäure = Glutamic acid

Glutaric acid **Glutarsäure**
sys[CA]: Pentanedioic acid

$C_5H_8O_4$

$HO-CO-[CH_2]_3-CO-OH$

Beilstein: 4-2.1934 · Syst.-Nr. 174
CAS-Reg.No.: 110-94-1
IUPAC: C-404.1

Glutinane **Glutinan**

= Oleanane, *D:B*-friedo-

Glyceric acid **Glycerinsäure**

sys[CA]: Propanoic acid, 2,3-dihydroxy-

$C_3H_6O_4$
Beilstein · Syst.-Nr. 230
CAS-Reg.No.: 473-81-4
IUPAC: C-411.1

$HO-CH_2-CH(OH)-CO-OH$

D-Glyceric acid
Beilstein: 4-3.1050
CAS-Reg.No.: 6000-40-4
L-Glyceric acid
Beilstein: 3-3.845
CAS-Reg.No.: 28305-26-2
DL-Glyceric acid
Beilstein: 4-3.1050
CAS-Reg.No.: 600-19-1

Glycerin = Glycerol

Glycerol **Glycerin**

sys[CA]: 1,2,3-Propanetriol

$C_3H_8O_3$
Beilstein: 4-1.2751 · Syst.-Nr. 38
CAS-Reg.No.: 56-81-5
IUPAC: C-201.4

$HO-CH_2-CH(OH)-CH_2-OH$

sn-Glycerol, formula A
IUPAC: Lip-1.13

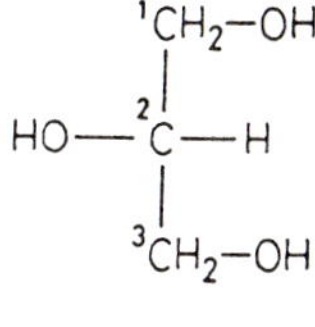

A

Glycine **Glycin**

$C_2H_5NO_2$
Beilstein: 4-4.2349 · Syst.-Nr. 364
CAS-Reg.No.: 56-40-6
IUPAC: 3AA-1

$H_2N-CH_2-CO-OH$

Glycolic acid **Glykolsäure**

sys[CA]: Acetic acid, hydroxy-

$C_2H_4O_3$ HO – CH$_2$ – CO – OH

Beilstein: 4-3.571 · Syst.-Nr. 220

CAS-Reg.No.: 79-14-1

IUPAC: C-411.1

Glyoxal **Glyoxal**

sys[CA]: Ethanedial

$C_2H_2O_2$ OCH – CHO

Beilstein: 4-1.3625 · Syst.-Nr. 95

CAS-Reg.No.: 107-22-2

IUPAC: C-305.3

Glyoxylic acid **Glyoxylsäure**

sys[CA]: Acetic acid, oxo-

$C_2H_2O_3$ OCH – CO – OH

Beilstein: 4-3.1489 · Syst.-Nr. 279

CAS-Reg.No.: 298-12-4

IUPAC: C-416.3

Gonane **Gonan**

$C_{17}H_{28}$

Beilstein: 3-5.1102 · Syst.-Nr. 472

CAS-Reg.No.: 4732-76-7

IUPAC: 2S-2.1(16)

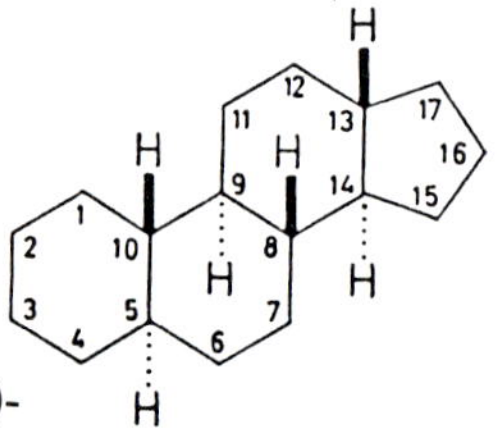

(5α)-

Gorgostane **Gorgostan**

$C_{30}H_{52}$

Beilstein: 4-6.4189* · Syst.-Nr. 473

CAS-Reg.No.: 36564-41-7

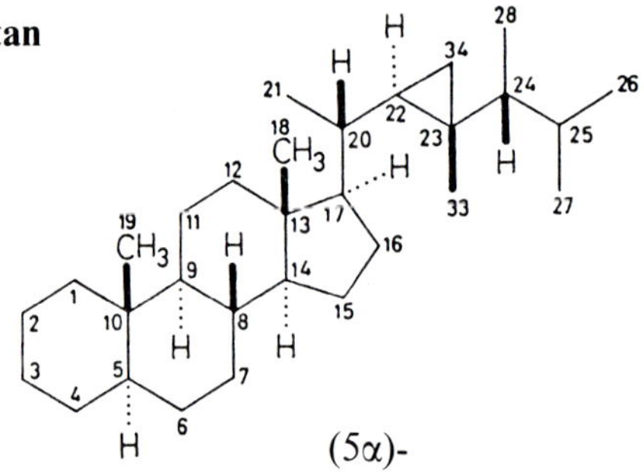

(5α)-

Grayanotoxane **Grayanotoxan**

$C_{20}H_{34}$

Beilstein · Syst.-Nr. 472

CAS-Reg.No.: 39907-73-8

IUPAC: F-2

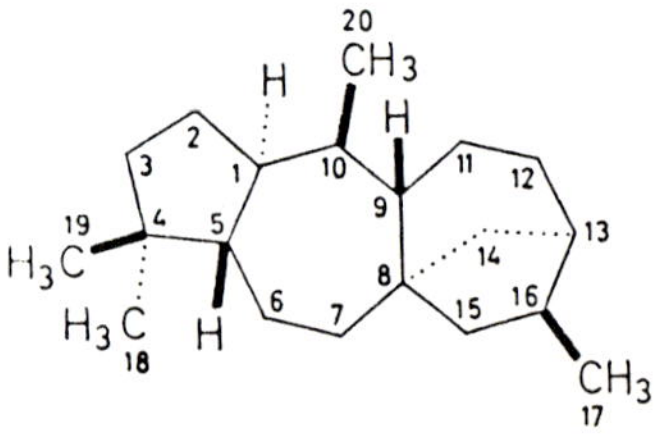

Grisan **Grisan**

= Spiro[benzofuran-2,1′-cyclohexane]

Guaiacol **Guajakol**

sys[CA]: Phenol, 2-methoxy-

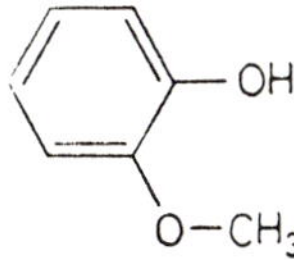

$C_7H_8O_2$
Beilstein: 4-6.5563 · Syst.-Nr. 553
CAS-Reg.No.: 90-05-1
IUPAC: C-214.1

Guaiane **Guajan**

sys[CA]: Azulene, decahydro-1,4-dimethyl-7-(1-methylethyl)-, [1*S*-(1α,3aβ,4α,7α,8aβ)]-

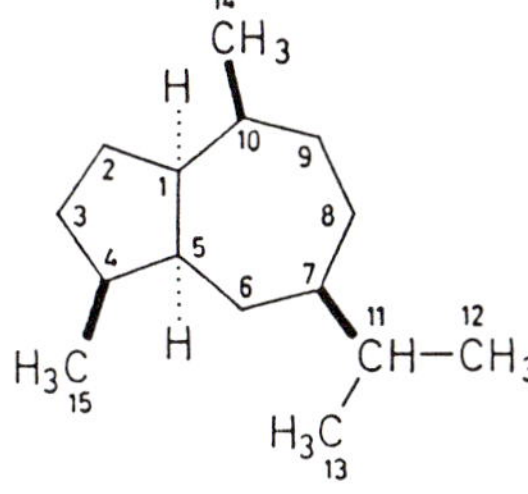

$C_{15}H_{28}$
Beilstein: 4-17.4677* · Syst.-Nr. 453
CAS-Reg.No.: 489-80-5

Guanidine **Guanidin**

CH_5N_3, formula A [IUPAC-numbering] and formula B [CA-numbering]
Beilstein: 4-3.148 · Syst.-Nr. 207
CAS-Reg.No.: 113-00-8
IUPAC: C-961.1

A B

Guanine **Guanin**

sys[B]: 6*H*-Purin-6-on, 2-Amino-1,7-dihydro-
[CA]: 6*H*-Purin-6-one, 2-amino-1,7-dihydro-

$C_5H_5N_5O$
Beilstein: 4-26.3890 · Syst.-Nr. 4179
CAS-Reg.No.: 73-40-5

Guanosine **Guanosin**

$C_{10}H_{13}N_5O_5$
Beilstein: 4-26.3901 · Syst.-Nr. 4179
CAS-Reg.No.: 118-00-3

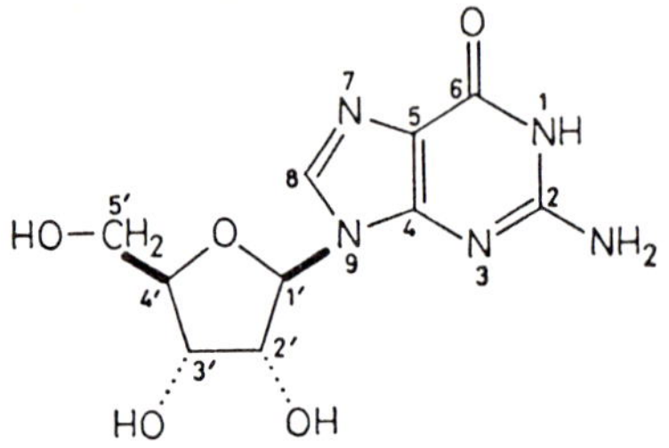

Gulofuranose **Gulofuranose**

$C_6H_{12}O_6$
Beilstein · Syst.-Nr. 144
IUPAC: Carb-5/18

α-D-Gulofuranose
CAS-Reg.No.: 36574-19-3
β-D-Gulofuranose
CAS-Reg.No.: 41847-50-1

α-L-Gulofuranose
CAS-Reg-No.: 40461-80-1
β-L-Gulofuranose
CAS-Reg.No.: 41847-48-7

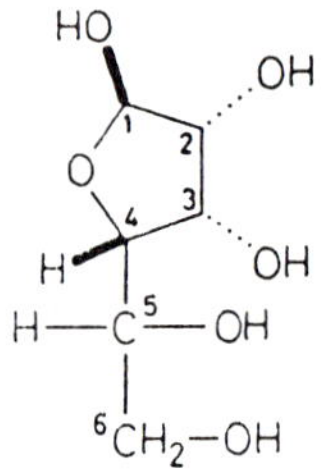

β-D-

Gulopyranose **Gulopyranose**

$C_6H_{12}O_6$
Beilstein · Syst.-Nr. 144
IUPAC: Carb-5/18

α-D-Gulopyranose
Beilstein: 4-1.4333
CAS-Reg.No.: 7282-78-2
β-D-Gulopyranose
Beilstein: 4-1.4333
CAS-Reg.No.: 7283-08-1
α-L-Gulopyranose
Beilstein: 4-1.4335
CAS-Reg.No.: 39281-66-8
β-L-Gulopyranose
CAS-Reg.No.: 39281-67-9

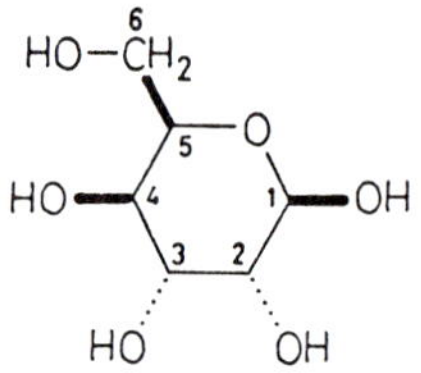

β-D-

Gulose **Gulose**

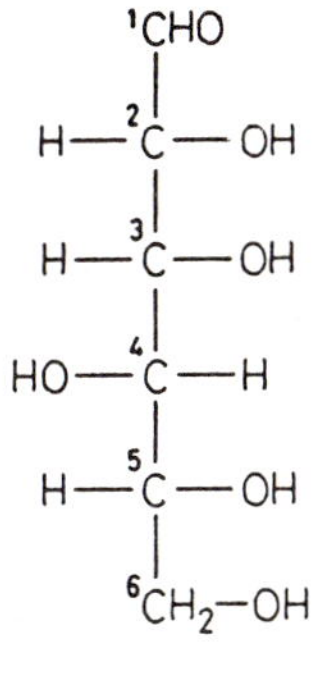

$C_6H_{12}O_6$
Beilstein · Syst.-Nr. 144
CAS-Reg.No.: 19163-87-2
IUPAC: Carb-5

D-Gulose
Beilstein: 4-1.4333
CAS-Reg.No.: 4205-23-6
–, 6-deoxy- = *Antiarose*

L-Gulose
Beilstein: 4-1.4334
CAS-Reg.No.: 6027-89-0

D-

Hämatoporphyrin = Hematoporphyrin

Hamamelose **Hamamelose**

= D-Ribose, 2-hydroxymethyl-

Harnstoff = Urea

Hasubanan **Hasubanan**

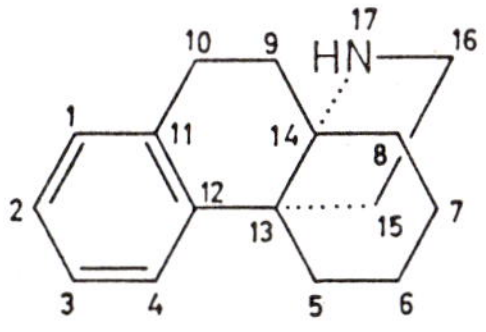

$C_{16}H_{21}N$
Beilstein: 4-21.1359* · Syst.-Nr. 3082
CAS-Reg.No.: 14510-67-9
IUPAC: F-2

Hematoporphyrin **Hämatoporphyrin**

sys[CA]: 21*H*,23*H*-Porphine-2,18-dipropanoic acid, 7,12-bis-(1-hydroxyethyl)-3,8,13,17-tetramethyl-

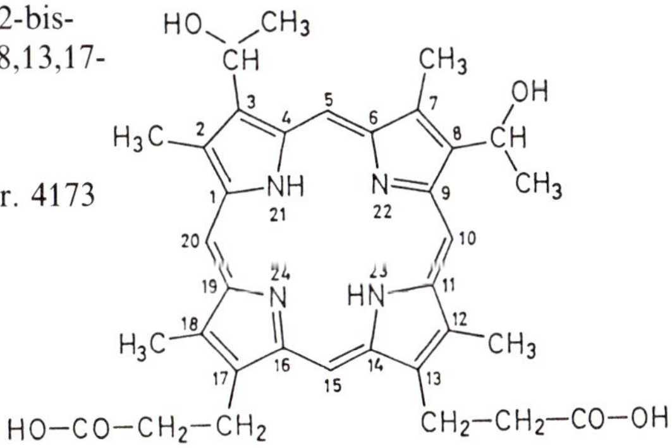

$C_{34}H_{38}N_4O_6$
Beilstein: 4-26.3157 · Syst.-Nr. 4173
CAS-Reg.No.: 14459-29-1
IUPAC: TP-2.1

Heptalene **Heptalen**

$C_{12}H_{10}$
Beilstein · Syst.-Nr. 479
CAS-Reg.No.: 257-24-9
IUPAC: A-21.2(5)

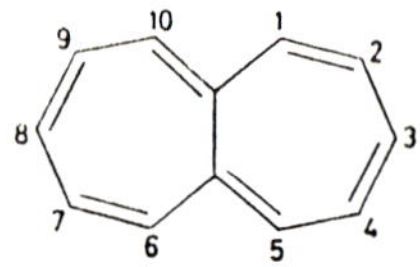

Heteratisane **Heteratisan**

$C_{18}H_{27}NO$
Beilstein: 4-27.3718 · Syst.-Nr. 4195
CAS-Reg.No.: 41855-38-3

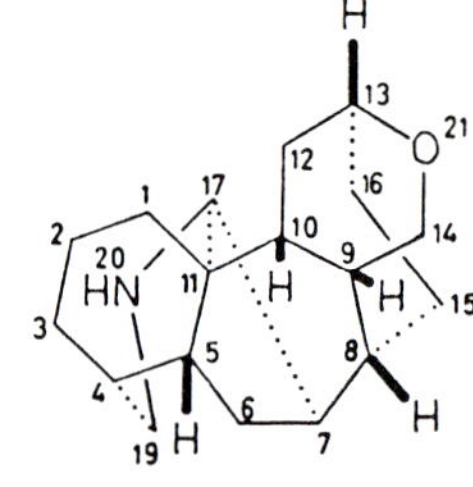

Hetisan **Hetisan**

$C_{20}H_{27}N$
Beilstein: 4-21.2308* · Syst.-Nr. 3083
CAS-Reg.No.: 39608-70-3
IUPAC: F-2

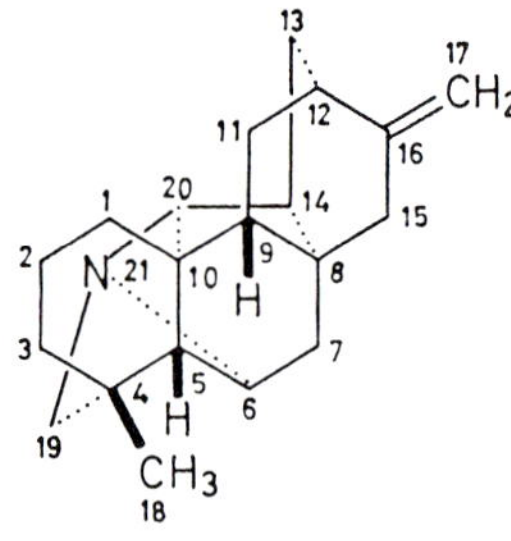

Himbosan **Himbosan**

$C_{21}H_{31}N$
Beilstein: 4-22.2616* · Syst.-Nr. 3082
CAS-Reg.No.: 41904-73-8
IUPAC: F-2

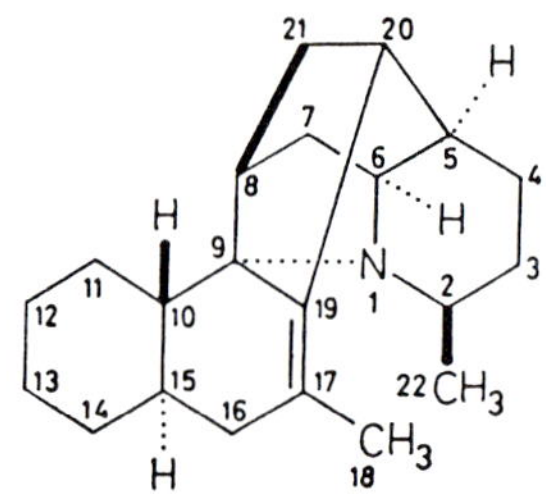

Hippuric acid **Hippursäure**

sys[B]: Glycin, *N*-Benzoyl-
[CA]: Glycine, *N*-benzoyl-

$C_9H_9NO_3$
Beilstein: 4-9.778 · Syst.-Nr. 920
CAS-Reg.No.: 495-69-2
IUPAC: C-421.4

—CO—NH—CH_2—CO—OH

Histidine **Histidin**

$C_6H_9N_3O_2$
Beilstein · Syst.-Nr. 3776
CAS-Reg.No.: 7006-35-1
IUPAC: 3AA-2.2.4

$CH_2-CH(NH_2)-CO-OH$

D-Histidine
Beilstein: 4-25.4343
CAS-Reg.No.: 351-50-8
L-Histidine
Beilstein: 4-25.4344
CAS-Reg.No.: 71-00-1
DL-Histidine
Beilstein: 4-25.4348
CAS-Reg.No.: 4998-57-6

Holostane **Holostan**

= Lanostane, 18,20-epoxy-(5α)-

Homocysteine **Homocystein**
sys[B]: Buttersäure, 2-Amino-4-mercapto-

$C_4H_9NO_2S$

$HS-CH_2-CH_2-CH(NH_2)-CO-OH$

Beilstein · Syst.-Nr. 376
CAS-Reg.No.: 454-28-4
IUPAC: C-421.2; 3AA-Appendix

D-Homocysteine
Beilstein: 3-4.1645
CAS-Reg.No.: 6027-14-1
L-Homocysteine
Beilstein: 3-4.1638
CAS-Reg.No.: 6027-13-0
DL-Homocysteine
Beilstein: 4-4.3189
CAS-Reg.No.: 454-29-5

Homoormosanine **Homoormosanin**

$C_{21}H_{35}N_3$
Beilstein: 4-26.166 · Syst.-Nr. 3808
CAS-Reg.No.: 10550-80-8

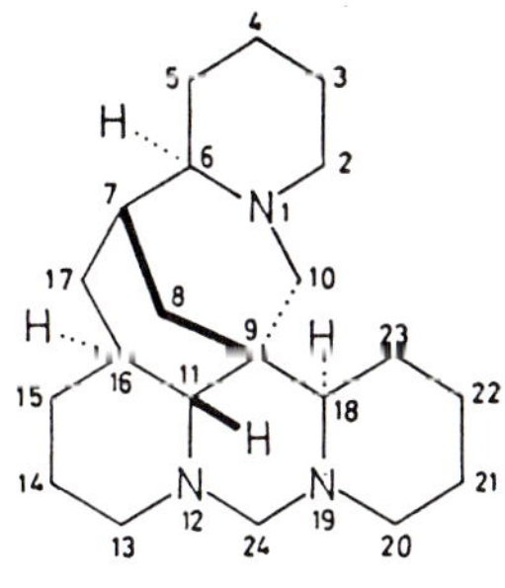

Homoserine **Homoserin**

sys[B]: Buttersäure, 2-Amino-4-hydroxy-

$C_4H_9NO_3$

$HO-CH_2-CH_2-CH(NH_2)-CO-OH$

Beilstein · Syst.-Nr. 376
CAS-Reg.No.: 498-19-1
IUPAC: C-421.1; 3AA-Appendix

D-Homoserine
Beilstein: 4-4.3187
CAS-Reg.No.: 6027-21-0
L-Homoserine
Beilstein: 4-4.3187
CAS-Reg.No.: 672-15-1
DL-Homoserine
Beilstein: 4-4.3187
CAS-Reg.No.: 1927-25-9

Hopane **Hopan**

= Gammacerane, *A'*-neo-

Hydantoic acid **Hydantoinsäure**

sys[B]: Glycin, *N*-Carbamoyl-
[CA]: Glycine, *N*-(aminocarbonyl)-

$C_3H_6N_2O_3$

$H_2N-CO-NH-CH_2-CO-OH$

Beilstein: 4-4.2411 · Syst.-Nr. 364
CAS-Reg.No.: 462-60-2
IUPAC: C-971.2

Hydantoin **Hydantoin**

sys[B]: Imidazolidin-2,4-dion
[CA]: 2,4-Imidazolidinedione

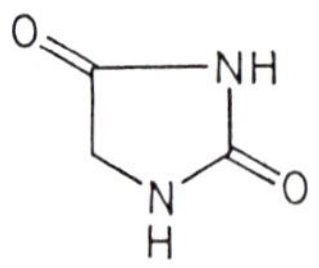

$C_3H_4N_2O_2$
Beilstein: 4-24.1034 · Syst.-Nr. 3587
CAS-Reg.No.: 461-72-3

Hydroquinone **Hydrochinon**

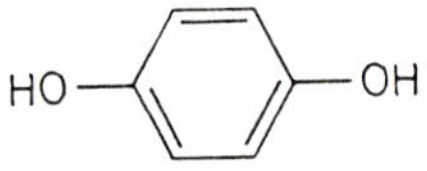

$C_6H_6O_2$
Beilstein: 4-6.5712 · Syst.-Nr. 555
CAS-Reg.No.: 123-31-9
IUPAC: C-202.2

Hypoxanthine **Hypoxanthin**

syn: Sarkin

sys[B]: Purin-6-on, 1,7-Dihydro-

[CA]: 6*H*-Purin-6-one, 1,7-dihydro-

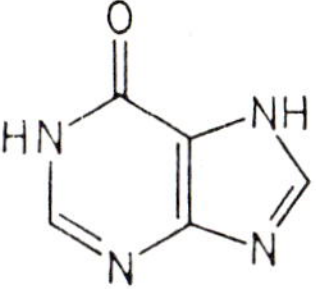

$C_5H_4N_4O$

Beilstein: 4-26.2081 · Syst.-Nr. 4115

CAS-Reg.No.: 68-94-0

Ibogamine **Ibogamin**

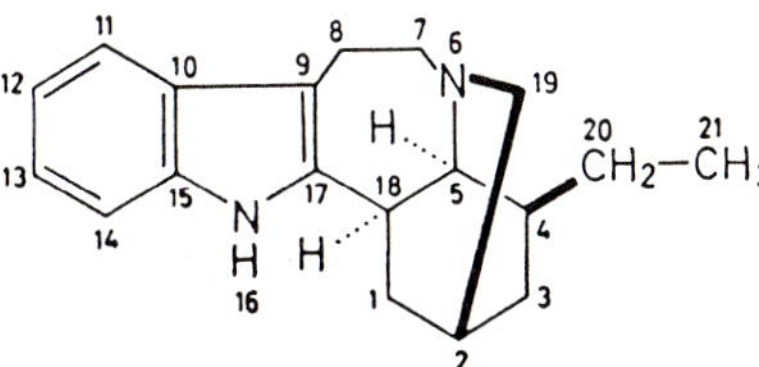

$C_{19}H_{24}N_2$

Beilstein: 4-23.1648 · Syst.-Nr. 3486

CAS-Reg.No.: 481-87-8

IUPAC: F-2

Idofuranose **Idofuranose**

$C_6H_{12}O_6$

Beilstein · Syst.-Nr. 144

IUPAC: Carb-5/18

α-D-Idofuranose

Beilstein: 4-1.4335

CAS-Reg.No.: 41487-67-0

β-D-Idofuranose

Beilstein: 4-1.4335

CAS-Reg.No.: 40461-75-4

α-L-Idofuranose

CAS-Reg.No.: 41847-65-8

β-L-Idofuranose

CAS-Reg.No.: 36574-15-9

β-D-

Idopyranose **Idopyranose**

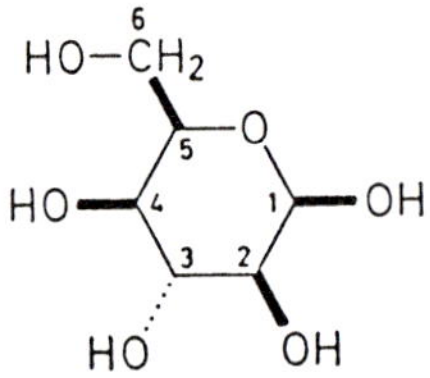

$C_6H_{12}O_6$
Beilstein · Syst.-Nr. 144
IUPAC: Carb-5/18

α-D-Idopyranose
Beilstein: 4-1.4335
CAS-Reg.No.: 7282-82-8
β-D-Idopyranose
Beilstein: 4-1.4335
CAS-Reg.No.: 7283-02-5
α-L-Idopyranose
CAS-Reg.No.: 39281-68-0
β-L-Idopyranose
CAS-Reg.No.: 12773-33-0

β-D-

Idose **Idose**

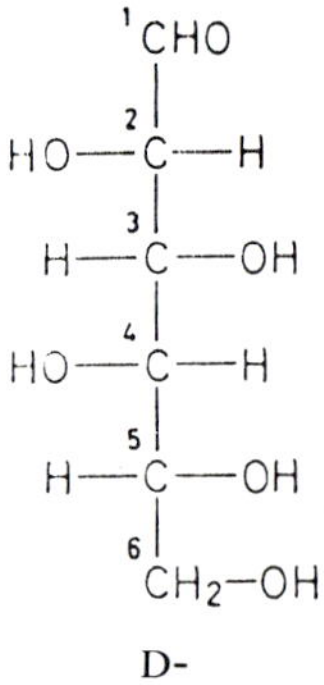

$C_6H_{12}O_6$
Beilstein · Syst.-Nr. 144
CAS-Reg.No.: 2152-76-3
IUPAC: Carb-5

D-Idose
Beilstein: 4-1.4335
CAS-Reg.No.: 5978-95-0
L-Idose
Beilstein: 4-1.4336
CAS-Reg.No.: 5934-56-5

D-

Imidazole **Imidazol**

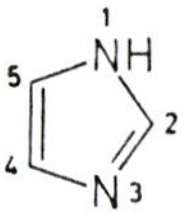

$C_3H_4N_2$
Beilstein: 4-23.564 · Syst.-Nr. 3463
IUPAC: B-2.11(13)

1*H*-Imidazole
CAS-Reg.No.: 288-32-4
2*H*-Imidazole
CAS-Reg.No.: 288-31-3
4*H*-Imidazole
CAS-Reg.No.: 288-30-2

Imidazolidine **Imidazolidin**

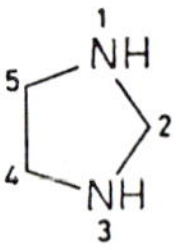

$C_3H_8N_2$
Beilstein: 4-23.5 · Syst.-Nr. 3460 A
CAS-Reg.No.: 504-74-5
IUPAC: B-2.12(5)

Imidazoline **Imidazolin**

sys[CA]: 1*H*-Imidazole, dihydro-

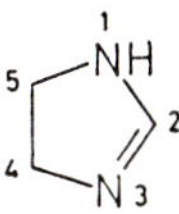

$C_3H_6N_2$
Beilstein: 4-23.455 · Syst.-Nr. 3461
CAS-Reg.No.: 28299-33-4
IUPAC: B-2.12(6)

***as*-Indacene** ***as*-Indacen**

$C_{12}H_8$
Beilstein · Syst.-Nr. 480
CAS-Reg.No.: 210-65-1
IUPAC: A-21.2(7)

***s*-Indacene** ***s*-Indacen**

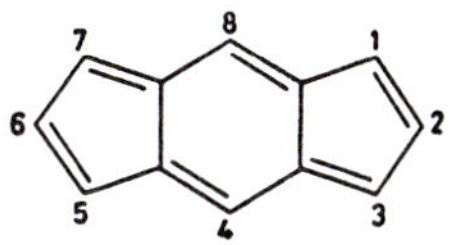

$C_{12}H_8$
Beilstein · Syst.-Nr. 480
CAS-Reg.No.: 267-21-0
IUPAC: A-21.2(8)

Indan **Indan**

sys[CA]: 1*H*-Indene, 2,3-dihydro-

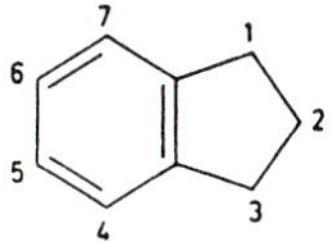

C_9H_{10}
Beilstein: 4-5.1371 · Syst.-Nr. 473
CAS-Reg.No.: 496-11-7
IUPAC: A-23.1

Indazole **Indazol**

sys[CA]: 1*H*-Indazole

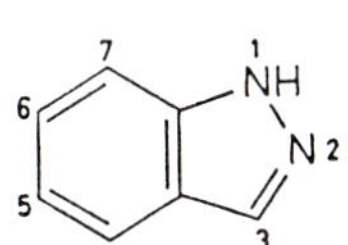

$C_7H_6N_2$
Beilstein: 4-23.1055 · Syst.-Nr. 3473
IUPAC: B-2.11(23)

1*H*-Indazole
CAS-Reg.No.: 271-44-3
2*H*-Indazole
CAS-Reg.No.: 271-42-1
3*H*-Indazole
CAS-Reg.No.: 271-43-2
3a*H*-Indazole
CAS-Reg.No.: 271-38-5

4*H*-Indazole
CAS-Reg.No.: 271-41-0
5*H*-Indazole
CAS-Reg.No.: 271-40-9
7*H*-Indazole
CAS-Reg.No.: 271-39-6

Indene **Inden**

sys[CA]: 1*H*-Indene

C_9H_8
Beilstein: 4-5.1532 · Syst.-Nr. 474
IUPAC: A-21.2(2)

1*H*-Indene
CAS-Reg.No.: 95-13-6
2*H*-Indene
CAS-Reg.No.: 270-53-1
3a*H*-Indene
CAS-Reg.No.: 270-51-9
5*H*-Indene
CAS-Reg.No.: 270-52-0

Indole **Indol**

sys[CA]: 1*H*-Indole

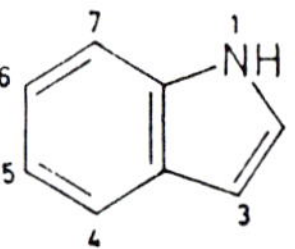

C_8H_7N
Beilstein: 4-20.3176 · Syst.-Nr. 3069
IUPAC: B-2.11(22)

1*H*-Indole
CAS-Reg.No.: 120-72-9
2*H*-Indole
CAS-Reg.No.: 271-25-0
3*H*-Indole
CAS-Reg.No.: 271-26-1
3a*H*-Indole
CAS-Reg.No.: 271-23-8
4*H*-Indole
CAS-Reg.No.: 271-24-9

Indoline **Indolin**

sys[CA]: 1*H*-Indole, 2,3-dihydro-

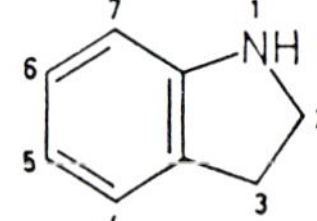

C_8H_9N
Beilstein: 4-20.2896 · Syst.-Nr. 3061
CAS-Reg.No.: 496-15-1
IUPAC: B-2.12(11)

Indolizine **Indolizin**

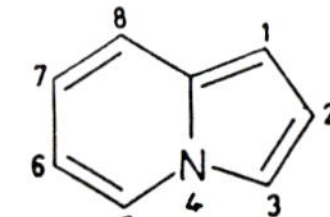

C_8H_7N
Beilstein: 4-20.3195 · Syst.-Nr. 3069
CAS-Reg.No.: 274-40-8
IUPAC: B-2.11(19)

Inosamine **Inosamin**

$C_6H_{13}NO_5$

Collective name for the amino-deoxy-inositols; individual compounds are given systematic names (IUPAC: I-6)

Beilstein: 4-13.3019 · Syst.-Nr. 1871

Inosine **Inosin**

$C_{10}H_{12}N_4O_5$

Beilstein: 4-26.2087 · Syst.-Nr. 4115

CAS-Reg.No.: 58-63-9

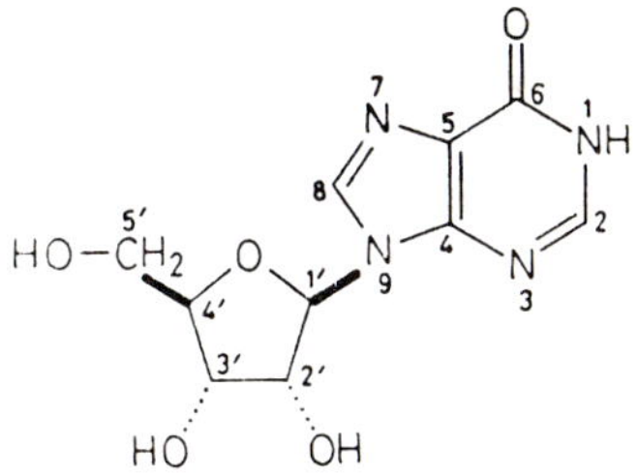

Inositol **Inosit**

$C_6H_{12}O_6$
Beilstein · Syst.-Nr. 604
CAS-Reg.No.: 6917-35-7
IUPAC: I-1.1

cis-Inositol, formula 1
Beilstein: 4-6.7919
CAS-Reg.No.: 576-63-6
epi-Inositol, formula 2
Beilstein: 4-6.7919
CAS-Reg.No.: 488-58-4
allo-Inositol, formula 3
Beilstein: 4-6.7919
CAS-Reg.No.: 643-10-7
myo-Inositol, formula 4
Beilstein: 4-6.7919
CAS-Reg.No.: 87-89-8
neo-Inositol, formula 5
Beilstein: 4-6.7919
CAS-Reg.No.: 488-54-0
muco-Inositol, formula 6
Beilstein: 4-6.7920
CAS-Reg.No.: 488-55-1
D-*chiro*-Inositol, formula 7
Beilstein: 4-6.7920
CAS-Reg.No.: 643-12-9
L-*chiro*-Inositol, formula *ent*-7
Beilstein: 4-6.7920
CAS-Reg.No.: 551-72-4
DL-*chiro*-Inositol, formula *rac*-7
Beilstein: 4-6.7920
CAS-Reg.No.: 18685-70-6
scyllo-Inositol, formula 8
Beilstein: 4-6.7920
CAS-Reg.No.: 488-59-5

HO OH HO OH HO OH
1

HO OH HO OH HO OH
2

HO OH HO OH HO OH
3

HO OH HO OH HO OH
4

HO OH HO OH HO OH
5

HO OH HO OH HO OH
6

HO OH HO OH HO OH
7

HO OH HO OH HO OH
8

Inosose **Inosose**

$C_6H_{10}O_6$
Collective name for the penta-hydroxy-cyclohexanones; individual compounds are given systematic names (IUPAC: I-8)
Beilstein: 4-8.3602 · Syst.-Nr. 843

Insularine **Insularin**

$C_{38}H_{40}N_2O_6$
Beilstein: 4-27.9041 · Syst.-Nr. 4660
CAS-Reg.No.: 549-07-5

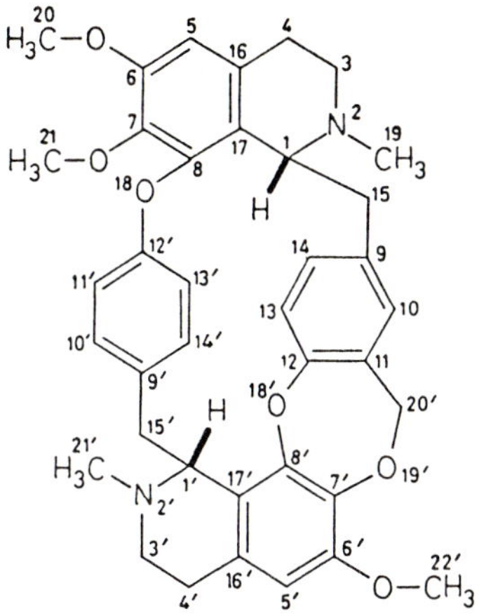

Iridane **Iridan**

syn: Osmane
sys[B]: Cyclopentan, 1-Isopropyl-2,3-dimethyl-
[CA]: Cyclopentane, 1,2-dimethyl-3-(1-methylethyl)-

$C_{10}H_{20}$
Beilstein: 4-6.166* · Syst.-Nr. 452
CAS-Reg.No.: 489-20-3

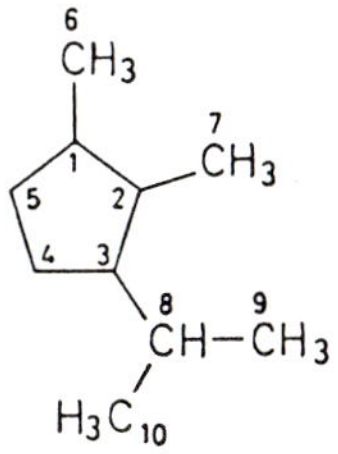

Isoatisine **Isoatisin**

$C_{22}H_{33}NO_2$
Beilstein: 4-27.1999 · Syst.-Nr. 4223
CAS-Reg.No.: 510-38-3

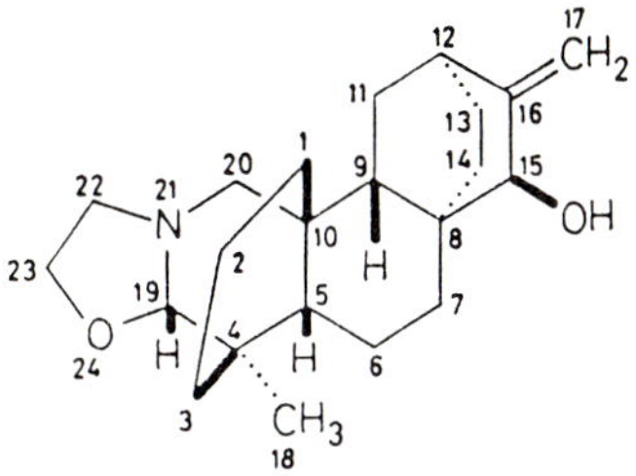

Isobenzofuran **Isobenzofuran**

C_8H_6O
Beilstein · Syst.-Nr. 2367
CAS-Reg.No.: 270-75-7
IUPAC: B-2.11(7)

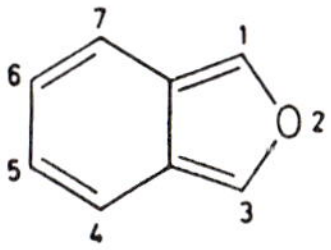

Isobutane **Isobutan**

sys[CA]: Propane, 2-methyl-

C_4H_{10} $(CH_3)_2CH-CH_3$

Beilstein: 4-1.282 · Syst.-Nr. 10

CAS-Reg.No.: 75-28-5

IUPAC: A-2.1

Isobutyric acid **Isobuttersäure**

= Propionic acid, 2-methyl-

Isocalciol **Isocalciol**

sys[CA]: 9,10-Secocholesta-1(10),5,7-trien-3-ol, (3β,5*E*,7*E*)-

$C_{27}H_{44}O$

Beilstein · Syst.-Nr. 535

CAS-Reg.No.: 42607-12-5

IUPAC: VD-5

(5*Z*)- (5*E*)-

Isocamphoric acid **Isocamphersäure**

= Camphoric acid, *trans*-

Isochinolin = Isoquinoline

Isochroman **Isochroman**

sys[CA]: 1*H*-2-Benzopyran, 3,4-dihydro-

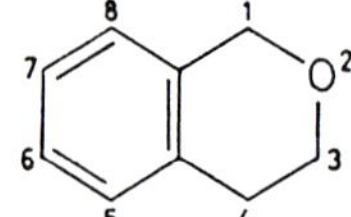

$C_9H_{10}O$

Beilstein: 5-17/1.595 · Syst.-Nr. 2366

CAS-Reg.No.: 493-05-0

IUPAC: B-2.12(1)

Isochromene **Isochromen**

sys[CA]: 1*H*-2-Benzopyran

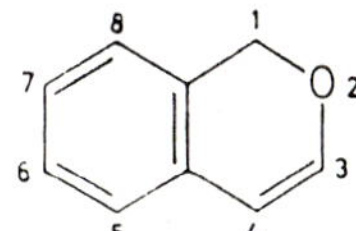

C_9H_8O
Beilstein: 5-17/2.21 · Syst.-Nr. 2367
CAS-Reg.No.: 253-35-0

Isocoumarine **Isocumarin**

sys[B]: Isochromen-1-on
[CA]: 1*H*-2-Benzopyran-1-one

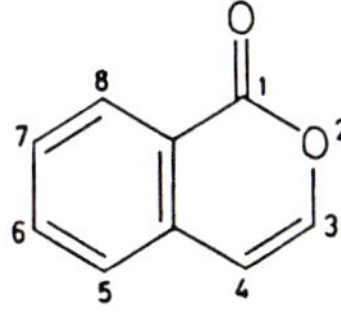

$C_9H_6O_2$
Beilstein: 4-17.5062 · Syst.-Nr. 2464
CAS-Reg.No.: 491-31-6
IUPAC: C-473.1

Isoflavan **Isoflavan**

= Chroman, 3-phenyl-

Isoflavene **Isoflaven**

= Chromene, 3-phenyl-

Isogenine **Isogenin**

= Spirostan, (25*R*)-

Isoharnstoff = Isourea

Isohexane **Isohexan**

sys[CA]: Pentane, 2-methyl-

C_6H_{14}

$(CH_3)_2CH-CH_2-CH_2-CH_3$

Beilstein: 4-1.358 · Syst.-Nr. 10
CAS-Reg.No.: 107-83-5
IUPAC: A-2.1

Isoindole **Isoindol**

sys[CA]: 2*H*-Isoindole

C_8H_7N
Beilstein · Syst.-Nr. 3069
IUPAC: B-2.11(20)

1*H*-Isoindole
CAS-Reg.No.: 270-69-9
2*H*-Isoindole
CAS-Reg.No.: 270-68-8

Isoindoline **Isoindolin**

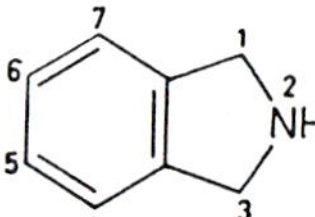

sys[CA]: 1*H*-Isoindole, 2,3-dihydro-

C_8H_9N
Beilstein: 4-20.2905 · Syst.-Nr. 3061
CAS-Reg.No.: 496-12-8
IUPAC: B-2.12(12)

Isoleontane **Isoleontan**

= Matridine, (6β)-

Isoleucine **Isoleucin**

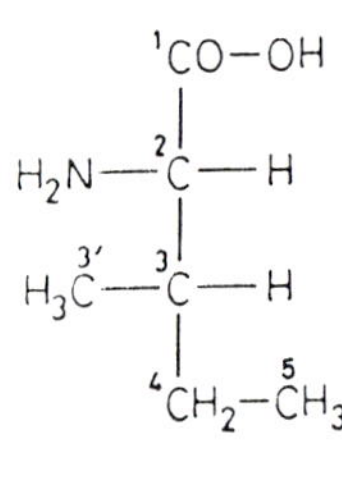

$C_6H_{13}NO_2$
Beilstein · Syst.-Nr. 368
CAS-Reg.No.: 7004-09-3
IUPAC: 3AA-2.2.1

D-Isoleucine
Beilstein: 4-4.2774
CAS-Reg.No.: 319-78-8
L-Isoleucine
Beilstein: 4-4.2775
CAS-Reg.No.: 73-32-5
DL-Isoleucine
Beilstein: 4-4.2776
CAS-Reg.No.: 443-79-8

Isomaltose **Isomaltose**

= D-Glucose, O^6-α-D-glucopyranosyl-

Isonicotinic acid **Isonicotinsäure**

sys[CA]: 4-Pyridinecarboxylic acid

$C_6H_5NO_2$
Beilstein: 4-22.518 · Syst.-Nr. 3249
CAS-Reg.No.: 55-22-1
IUPAC: C-404.1

Isopentane **Isopentan**

sys[CA]: Butane, 2-methyl-

C_5H_{12}
Beilstein: 4-1.320 · Syst.-Nr. 10
CAS-Reg.No.: 78-78-4
IUPAC: A-2.1

$(CH_3)_2CH-CH_2-CH_3$

Isophthalic acid **Isophthalsäure**

sys[CA]: 1,3-Benzenedicarboxylic acid

$C_8H_6O_4$
Beilstein: 4-9.3292 · Syst.-Nr. 977
CAS-Reg.No.: 121-91-5
IUPAC: C-404.1

Isoprene **Isopren**

sys[B]: Buta-1,3-dien, 2-methyl-
[CA]: 1,3-Butadiene, 2-methyl-

C_5H_8
Beilstein: 4-1.1001 · Syst.-Nr. 12
CAS-Reg.No.: 78-79-5
IUPAC: A-3.4

$CH_2=CH-C(CH_3)=CH_2$

Isoprimverose **Isoprimverose**

= D-Glucose, O^6-α-D-xylopyranosyl-

Isoquinoline **Isochinolin**

C_9H_7N
Beilstein: 4-20.3410 · Syst.-Nr. 3078
CAS-Reg.No.: 119-65-3
IUPAC: B-2.11(26)

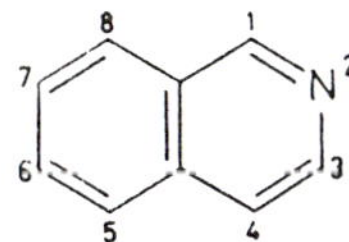

Isosafrole **Isosafrol**

sys[B]: Benzo[1,3]dioxol, 5-Propenyl-
[CA]: 1,3-Benzodioxole, 5-(1-propenyl)-

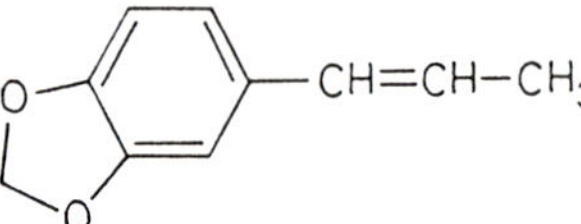

$C_{10}H_{10}O_2$
Beilstein: 4-19.273 · Syst.-Nr. 2673
CAS-Reg.No.: 120-58-1
IUPAC: C-331.3

Isothiazole **Isothiazol**

C_3H_3NS
Beilstein: 4-27.960 · Syst.-Nr. 4192
CAS-Reg.No.: 288-16-4
IUPAC: B-2.11(43)

Isourea **Isoharnstoff**

sys[CA]: Carbamimidic acid

CH_4N_2O
Beilstein · Syst.-Nr. 205 B
CAS-Reg.No.: 4744-36-9
IUPAC: C-972.1

$HN{=}C(OH){-}NH_2$

Isovaleric acid **Isovaleriansäure**

= Butyric acid, 3-methyl-

Isovaline **Isovalin**

$C_5H_{11}NO_2$
Beilstein · Syst.-Nr. 367
CAS-Reg.No.: 465-58-7

D-Isovaline
Beilstein: 4-4.2655
CAS-Reg.No.: 3059-97-0

L-Isovaline
Beilstein: 4-4.2655
CAS-Reg.No.: 595-40-4

DL-Isovaline
Beilstein: 4-4.2655
CAS-Reg.No.: 595-39-1

L-

Isoviolanthrene **Isoviolanthren**

sys[CA]: Benzo[*rst*]phenanthro[10,1,2-*cde*]pentaphene, 9,18-dihydro-

$C_{34}H_{20}$
Beilstein · Syst.-Nr. 497
CAS-Reg.No.: 4430-29-6
IUPAC: A-23.1

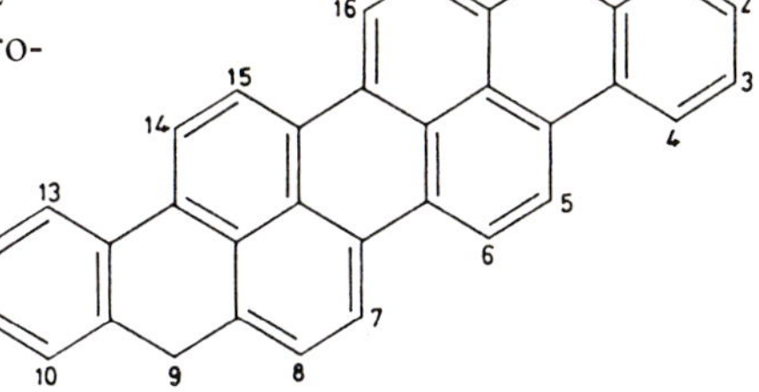

Isovitamin D_3 **Isovitamin-D_3**

= Isocalciol, (5*E*)-

Isoxazole **Isoxazol**

C_3H_3NO
Beilstein: 4-27.959 · Syst.-Nr. 4192
CAS-Reg.No.: 288-14-2
IUPAC: B-2.11(45)

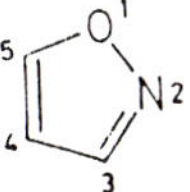

Isoxazolidine **Isoxazolidin**

C_3H_7NO
Beilstein: 4-27.6 · Syst.-Nr. 4190 A
CAS-Reg.No.: 504-72-3

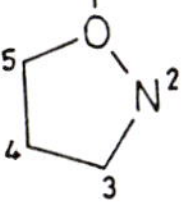

Jervane **Jervan**

sys[B+CA]: 12,14-Cyclo-13,14-seco-cholestane

$C_{27}H_{48}$
Beilstein · Syst.-Nr. 472
IUPAC: 2S-11.1(107)

(5α)-

Jervanine **Jervanin**

sys[B]: Veratran, 17,23-Epoxy-
sys[CA]: Veratraman, 17,23-epoxy-5,6,12,13-tetrahydro-

$C_{27}H_{45}NO$
Beilstein · Syst.-Nr. 4195
IUPAC: 2S-11.1(106)

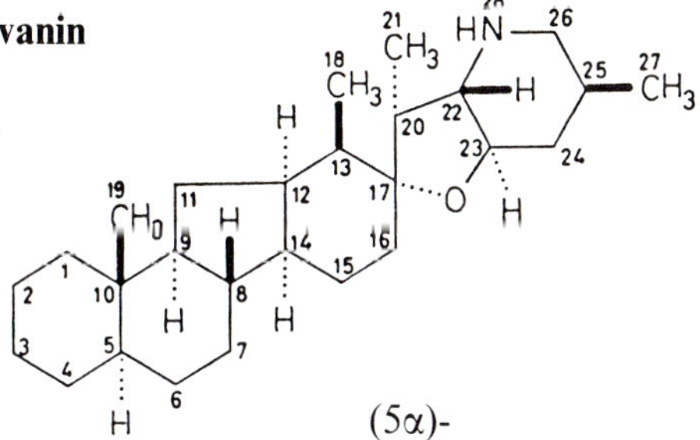

(5α)-

Kaurane **Kauran**

$C_{20}H_{34}$
Beilstein: 3-10.942* · Syst.-Nr. 472
CAS-Reg.No.: 1573-40-6

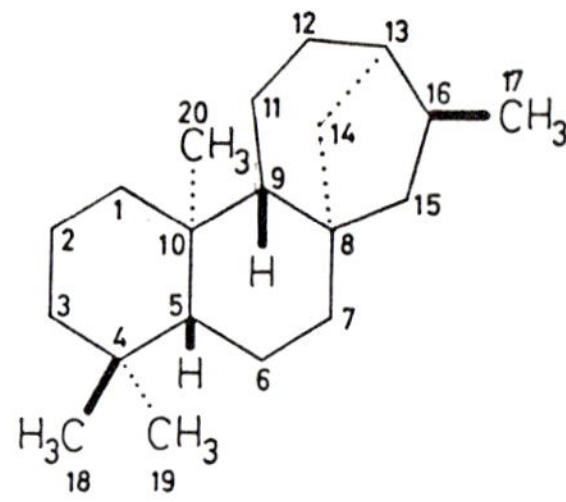

Ketene **Keten**

sys[CA]: Ethenone

C_2H_2O
Beilstein: 4-1.3418 · Syst.-Nr. 90
CAS-Reg.No.: 463-51-4
IUPAC: C-321.1

$CH_2=C=O$

Kohlensäure = Carbonic acid

Kojibiose **Kojibiose**

= D-Glucose, O^2-α-D-glucopyranosyl-

Kopsan **Kopsan**

$C_{20}H_{24}N_2$
Beilstein: 4-25.204 · Syst.-Nr. 3487
CAS-Reg.No.: 464-60-8

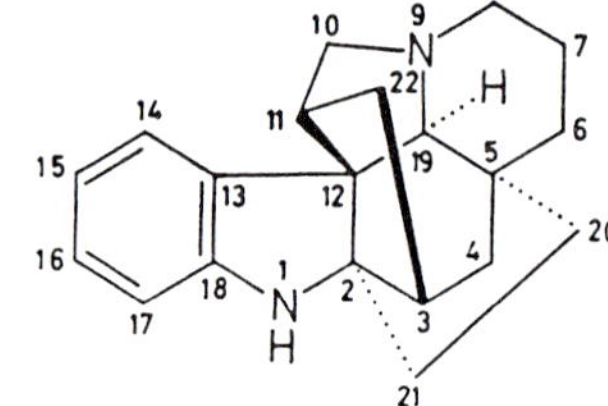

Koproporphyrin-I = Coproporphyrin I

Koprostan = Coprostane

Korksäure = Suberic acid

Kresol = Cresol

Labdane **Labdan**

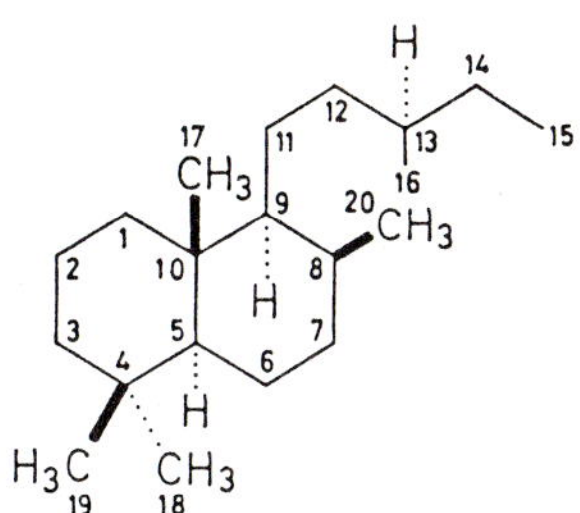

$C_{20}H_{38}$
Beilstein: 4-5.368 · Syst.-Nr. 453
CAS-Reg.No.: 561-90-0
IUPAC: F-(18)

–, (13*S*)-8,13-epoxy-*A*-nor- = *Colensane*

Lactic acid **Milchsäure**

sys[CA]: Propanoic acid, 2-hydroxy-

$C_3H_6O_3$
Beilstein: 4-3.633 · Syst.-Nr. 221
CAS-Reg.No.: 50-21-5
IUPAC: C-411.1

$H_3C-CH(OH)-CO-OH$

D-Lactic acid
CAS-Reg.No.: 10326-41-7
L-Lactic acid
CAS-Reg.No.: 79-33-4
DL-Lactic acid
CAS-Reg.No.: 598-82-3

Lactoflavin **Lactoflavin**

= Riboflavin

Lactose **Lactose**

= D-Glucose, O^4-β-D-galactopyranosyl-

Laminaribiose **Laminaribiose**

= D-Glucose, O^3-β-D-glucopyranosyl-

Lanostane **Lanostan**

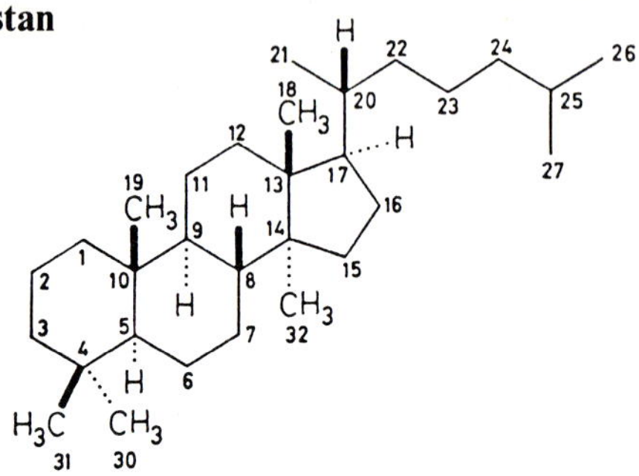

$C_{30}H_{54}$
Beilstein: 4-5.1241 · Syst.-Nr. 472
CAS-Reg.No.: 474-20-4
IUPAC: 2S-2.3(27)

–, (13α,14β,17β*H*)- = Euphane
–, *ent*-(5β,8α,9β,10α)- = Tirucallane
–, (5α)-9β,19-cyclo- = *Artostane*, *Cycloartane*
–, 18,20-epoxy-(5α)- = *Holostane*
–, (24*S*)-24-methyl- = Eburicane, *Laudane*
–, (24*R*)-24-methyl-(13α,14β,17β*H*,20β_F*H*)- = Euphorbane
–, 8-methyl-18-nor- = Dammarane
–, *ent*-8-methyl-18-nor-(5β,10α,20β_F*H*)- = *Protostane*
–, 9-methyl-19-nor-(9β)- = Cucurbitane
–, 32-nor- = Cholestane, 4,4-dimethyl-

Lanosterol **Lanosterin**

= Lanosta-8,24-dien-3β-ol

Lanthionine **Lanthionin**

sys[B]: Sulfid, Bis-[2-amino-2-carboxy-äthyl]-
[CA]: Cystein, *S*-(2-amino-2-carboxyethyl)-

$C_6H_{12}N_2O_4S$

$$HO-CO-CH(NH_2)-CH_2-S-CH_2-CH(NH_2)-CO-OH$$

Beilstein · Syst.-Nr. 376
CAS-Reg.No.: 496-98-0
IUPAC: C-421.1; 3AA-Appendix

D-Lanthionine
Beilstein: 3-4.1618
CAS-Reg.No.: 5965-92-4
L-Lanthionine
Beilstein: 3-4.1593
CAS-Reg.No.: 922-55-4
DL-Lanthionine
Beilstein: 3-4.1620
CAS-Reg.No.: 3183-08-2
meso-Lanthionine
Beilstein: 4-4.3152
CAS-Reg.No.: 922-56-5

Laudane **Laudan**

= Eburicane

Lauric acid **Laurinsäure**

= Dodecanoic acid

Leucine **Leucin**

$C_6H_{13}NO_2$
Beilstein · Syst.-Nr. 368
CAS-Reg.No.: 7005-03-0
IUPAC: 3AA-2.2.1

```
 5
H3C
   \4   3       2         1
    CH-CH2-CH(NH2)-CO-OH
   /
H3C
 5'
```

D-Leucine
Beilstein: 4-4.2738
CAS-Reg.No.: 328-38-1
L-Leucine
Beilstein: 4-4.2738
CAS-Reg.No.: 61-90-5
DL-Leucine
Beilstein: 4-4.2739
CAS-Reg.No.: 328-39-2

Leuconolide **Leuconolid**

$C_{20}H_{32}O_7$
Beilstein · Syst.-Nr. 2569
CAS-Reg.No.: 61828-46-4

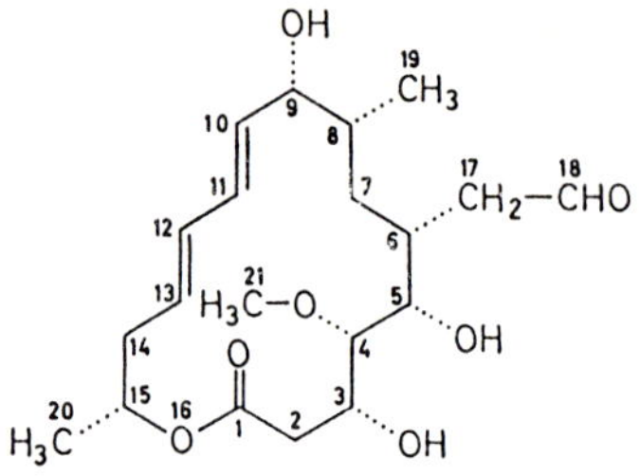

Limonoic acid **Limonsäure**

$C_{26}H_{34}O_{10}$
Beilstein · Syst.-Nr. 2986
CAS-Reg.No.: 22153-41-9

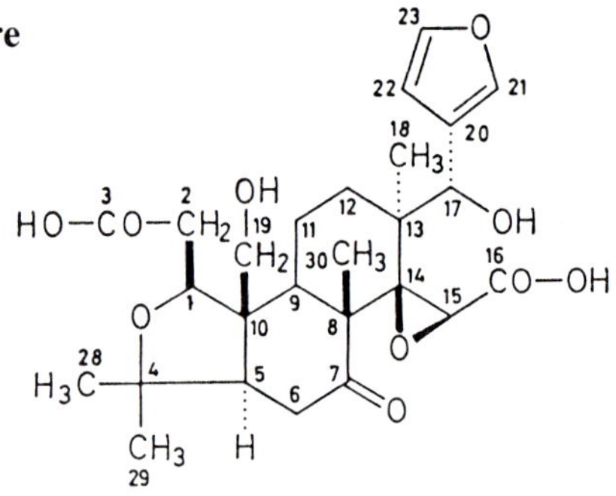

Lithocholic acid **Lithocholsäure**

= Cholan-24-oic acid, 3α-hydroxy-(5β)-

Lobane **Loban**

sys[CA]: Cyclohexane,
4-(1,5-dimethylhexyl)-1-ethyl-
1-methyl-2-(1-methylethyl)-

$C_{20}H_{40}$
Beilstein: 4-17.675* · Syst.-Nr. 452
CAS-Reg.No.: 71593-03-8

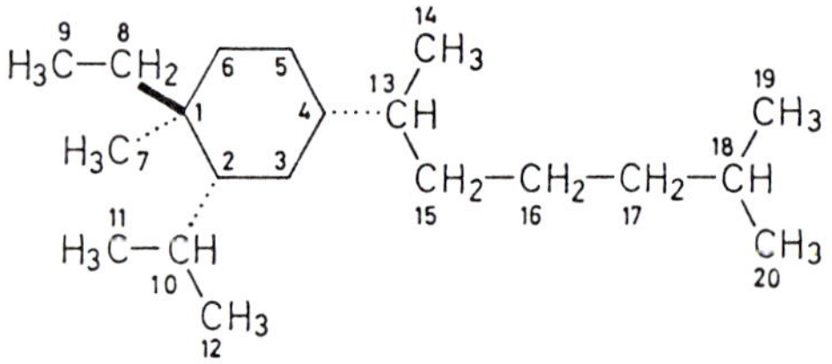

Longifolane **Longifolan**

sys[CA]: 1,4-Methanoazulene,
decahydro-4,8,8,9-tetramethyl-,
[1*R*-(1α,3aβ,4α,8aβ,9*S**)]-

$C_{15}H_{26}$
Beilstein: 4-5.498 · Syst.-Nr. 461
CAS-Reg.No.: 475-19-4

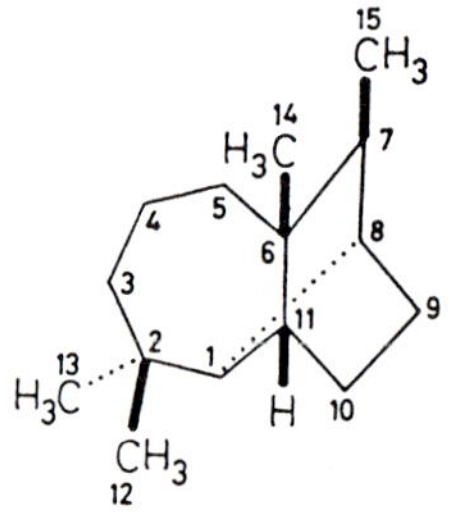

Lumistane **Lumistan**

sys[CA]: Ergostane, (5α,9β,10α)-

$C_{28}H_{50}$
Beilstein: 3-5.1927* · Syst.-Nr. 472
CAS-Reg.No.: 6538-00-7

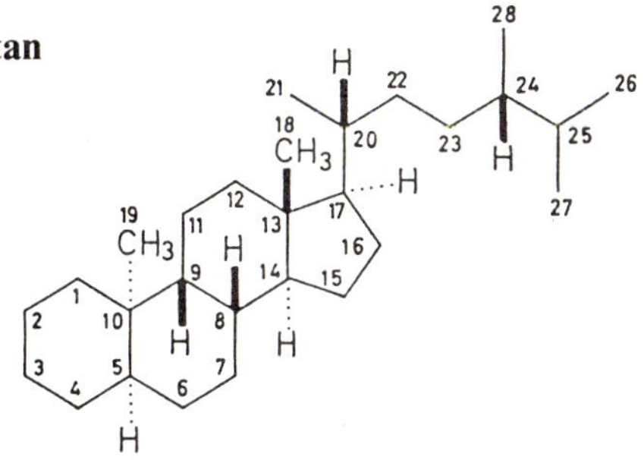

Lunaridine **Lunaridin**

$C_{25}H_{31}N_3O_4$
Beilstein: 4-27.9416 · Syst.-Nr. 4673
CAS-Reg.No.: 34340-56-2

(±)-Lunaridine
CAS-Reg.No.: 79298-93-4

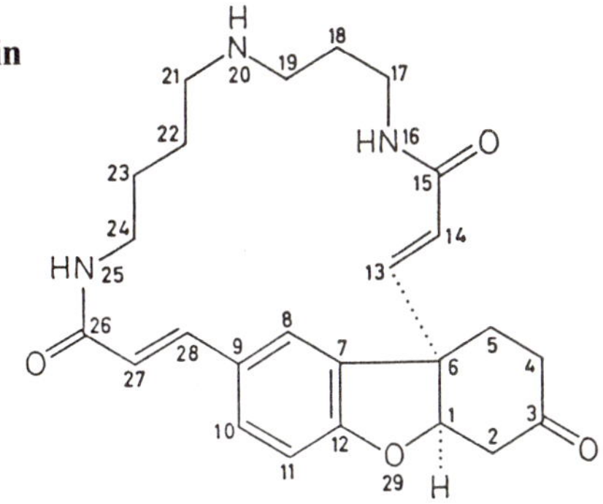

Lunarine **Lunarin**

$C_{25}H_{31}N_3O_4$
Beilstein: 4-27.9415 · Syst.-Nr. 4673
CAS-Reg.No.: 24185-51-1

(±)-Lunarine
CAS-Reg.No.: 79298-94-5

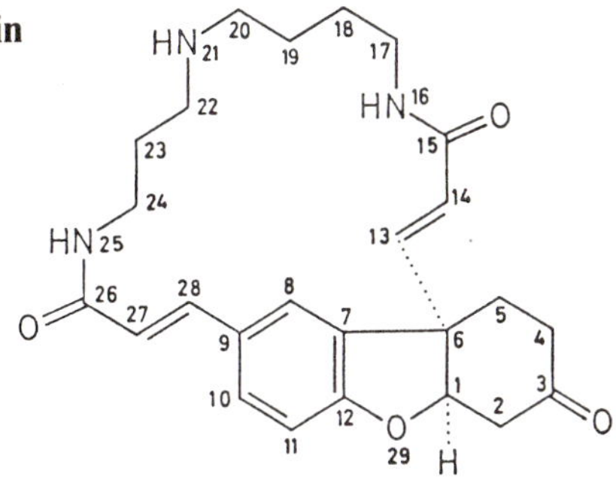

Lupane **Lupan**

$C_{30}H_{52}$
Beilstein: 3-5.1342 · Syst.-Nr. 473
CAS-Reg.No.: 464-99-3

–, *D:A*-friedo-18,19-seco- = *Shionane*

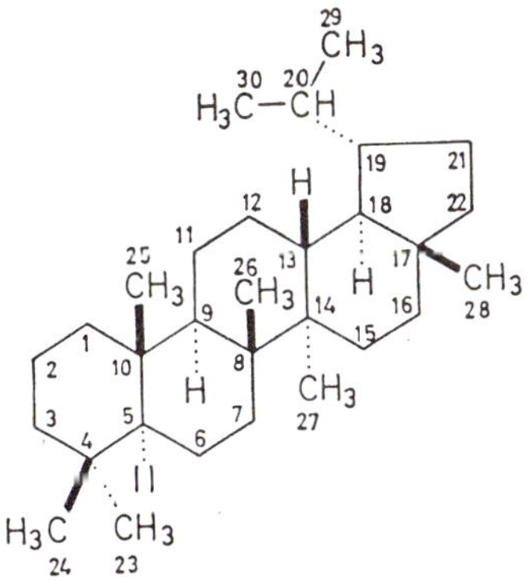

Lycobiose **Lycobiose**

= D-Galactose, O^4-β-D-glucopyranosyl-

Lycodine **Lycodin**

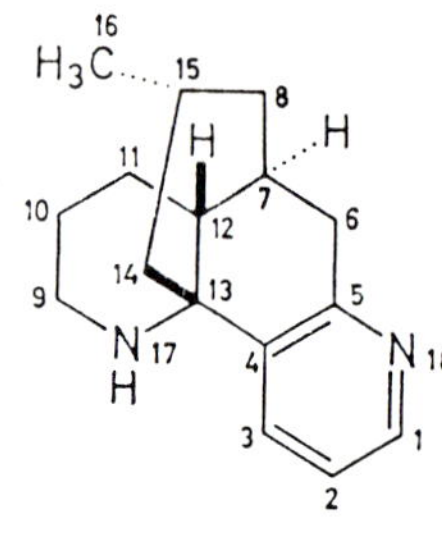

$C_{16}H_{22}N_2$
Beilstein: 4-23.1346 · Syst.-Nr. 3484
CAS-Reg.No.: 20316-18-1

Lycopene **Lycopin**

= ψ,ψ-Carotene

Lycopodane **Lycopodan**

$C_{15}H_{25}N$, formula A [Beilstein-numbering and formula B [CA-numbering]
Beilstein: 4-20.3081* · Syst.-Nr. 3056
CAS-Reg.No.: 34625-83-7
IUPAC: F-2

A

B

Lycoranan **Lycoranan**

sys[CA]: Lycorenan, 4,12-dihydro-

$C_{15}H_{19}NO$
Beilstein: 4-21.2542* · Syst.-Nr. 4196
IUPAC: F-2

Lycorenan **Lycorenan**

= Lycoran-4(12)-ene

Lysine **Lysin**

$C_6H_{14}N_2O_2$

$H_2N-[CH_2]_4-CH(NH_2)-CO-OH$

Beilstein: 4-4.2717 · Syst.-Nr. 368
CAS-Reg.No.: 6899-06-5
IUPAC: 3AA-1

D-Lysine
Beilstein: 4-4.2717
CAS-Reg.No.: 923-27-3
L-Lysine
Beilstein: 4-4.2717
CAS-Reg.No.: 56-87-1
DL-Lysine
Beilstein: 4-4.2720
CAS-Reg.No.: 70-54-2

Lythran **Lythran**

$C_{24}H_{27}NO$
Beilstein · Syst.-Nr. 4200
CAS-Reg.No.: 34620-86-5

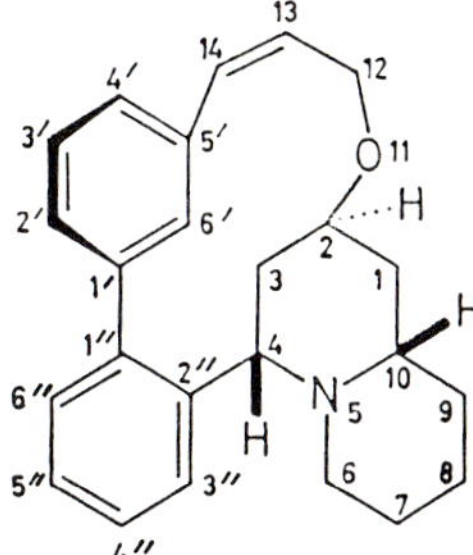

Lyxofuranose **Lyxofuranose**

$C_5H_{10}O_5$
Beilstein · Syst.-Nr. 133
IUPAC: Carb-5/18

HO−CH₂ O OH
HO OH

α-D-

α-D-Lyxofuranose
Beilstein: 4-1.4230
CAS-Reg.No.: 25545-04-4
β-D-Lyxofuranose
Beilstein: 4-1.4230
CAS-Reg.No.: 7687-39-0
α-L-Lyxofuranose
CAS-Reg.No.: 40461-89-0
β-L-Lyxofuranose
CAS-Reg.No.: 40461-77-6

Lyxopyranose **Lyxopyranose**

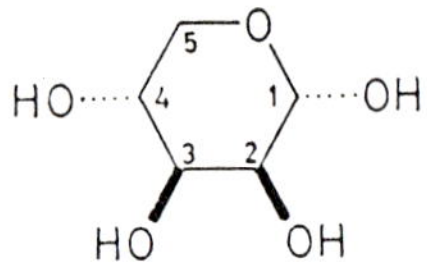

$C_5H_{10}O_5$
Beilstein · Syst.-Nr. 133
IUPAC: Carb-5/18

α-D-Lyxopyranose
Beilstein: 4-1.4230
CAS-Reg.No.: 608-46-8
β-D-Lyxopyranose
Beilstein: 4-1.4230
CAS-Reg.No.: 608-47-9
α-L-Lyxopyranose
Beilstein: 4-1.4232
CAS-Reg.No.: 7283-06-9
β-L-Lyxopyranose
Beilstein: 4-1.4232
CAS-Reg.No.: 7283-07-0

α-D-

Lyxose **Lyxose**

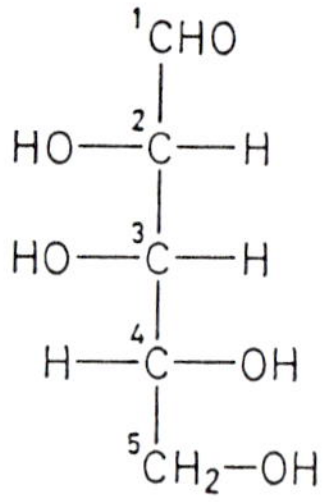

$C_5H_{10}O_5$
Beilstein · Syst.-Nr. 133
CAS-Reg.No.: 65-42-9
IUPAC: Carb-5

D-Lyxose
Beilstein: 4-1.4230
CAS-Reg.No.: 1114-34-7
L-Lyxose
Beilstein: 4-1.4232
CAS-Reg.No.: 1949-78-6

D-

Maleic acid **Maleinsäure**
sys[CA]: 2-Butenedioic acid, (*Z*)-

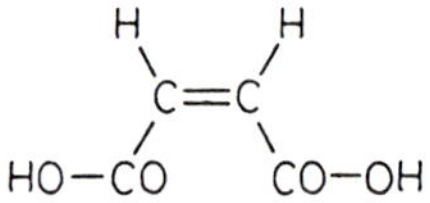

$C_4H_4O_4$
Beilstein: 4-2.2199 · Syst.-Nr. 179
CAS-Reg.No.: 110-16-7
IUPAC: C-404.1

–, methyl- = *Citraconic acid*

Malic acid **Äpfelsäure**

sys[B]: Bernsteinsäure, Hydroxy-
[CA]: Butanedioic acid, hydroxy-

$C_4H_6O_5$

$HO-CO-CH_2-CH(OH)-CO-OH$

Beilstein · Syst.-Nr. 240
CAS-Reg.No.: 6915-15-7
IUPAC: C-411.1

D-Malic acid
Beilstein: 4-3.1123
CAS-Reg.No.: 636-61-3
L-Malic acid
Beilstein: 4-3.1123
CAS-Reg.No.: 97-67-6
DL-Malic acid
Beilstein: 4-3.1124
CAS-Reg.No.: 617-48-1

Malonic acid **Malonsäure**

sys[CA]: Propanedioic acid

$C_3H_4O_4$

$HO\text{-}CO\text{-}CH_2\text{-}CO\text{-}OH$

Beilstein: 4-2.1874 · Syst.-Nr. 171
CAS-Reg.No.: 141-82-2
IUPAC: C-404.1

Maltose **Maltose**

= D-Glucose, O^4-α-D-glucopyranosyl-

Mannobiose **Mannobiose**

= D-Mannose,
O^4-β-D-mannopyranosyl-

Mannofuranose **Mannofuranose**

$C_6H_{12}O_6$
Beilstein · Syst.-Nr. 144
IUPAC: Carb-5/18

α-D-Mannofuranose
CAS-Reg.No.: 36574-21-7
β-D-Mannofuranose
Beilstein: 4-1.4329
CAS-Reg.No.: 40550-49-0
α-L-Mannofuranose
CAS-Reg.No.: 36972-23-3
β-L-Mannofuranose
CAS-Reg.No.: 37738-80-0

β-D-

Mannopyranose **Mannopyranose**

$C_6H_{12}O_6$
Beilstein · Syst.-Nr. 144
IUPAC: Carb-5/18

α-D-Mannopyranose
Beilstein: 4-1.4328
CAS-Reg.No.: 7296-15-3
β-D-Mannopyranose
Beilstein: 4-1.4328
CAS-Reg.No.: 7322-31-8
α-L-Mannopyranose
CAS-Reg.No.: 35810-56-1
β-L-Mannopyranose
Beilstein: 4-1.4333
CAS-Reg.No.: 12773-34-1

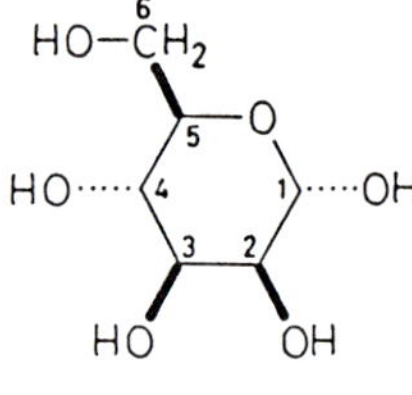

α-D-

Mannose **Mannose**

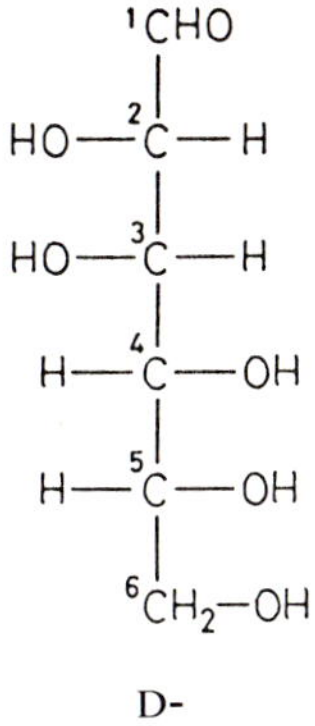

$C_6H_{12}O_6$
Beilstein · Syst.-Nr. 144
CAS-Reg.No.: 31103-86-3
IUPAC: Carb-5

–, 6-deoxy- = *Rhamnose*
–, O^3-methyl-6-deoxy- = *Acofriose*

D-Mannose
Beilstein: 4-1.4328
CAS-Reg.No.: 3458-28-4
–, O^4-β-D-galactopyranosyl- = *Epilactose*
–, O^6-α-D-galactopyranosyl- = *Epimelibiose*
–, O^4-β-D-glucopyranosyl- = *Epicellobiose*
–, O^6-β-D-glucopyranosyl- = *Epigentiobiose*
–, O^4-β-D-mannopyranosyl- = *Mannobiose*

L-Mannose
Beilstein: 4-1.4333
CAS-Reg.No.: 10030-80-5
–, O^4-β-D-glucopyranosyl-6-deoxy- = *Scillabiose*

D-

Matridine **Matridin**

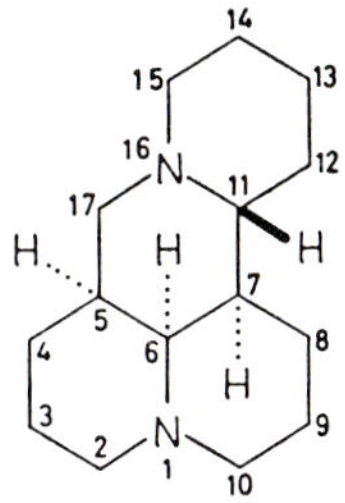

$C_{15}H_{26}N_2$
Beilstein: 4-23.963 · Syst.-Nr. 3469
CAS-Reg.No.: 569-24-4

–, (5β)- = *Sophoridane*
–, (6β)- = *Allomatridine*, *Isoleontane*

Maytansine **Maytansin**

$C_{34}H_{46}ClN_3O_{10}$
Beilstein · Syst.-Nr. 4642
CAS-Reg.No.: 35846-35-8

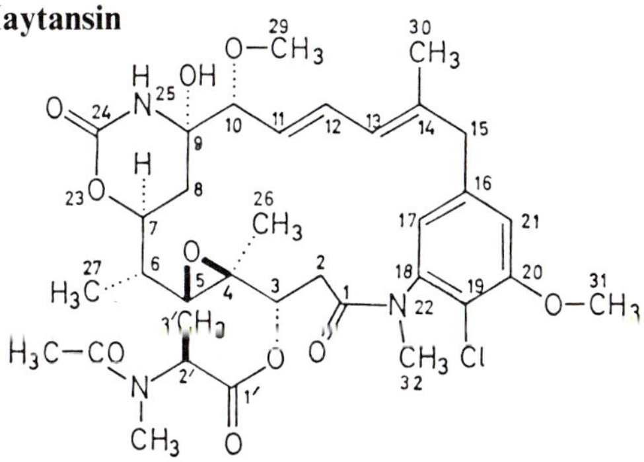

Melibiose **Melibiose**

= D-Glucose, O^6-α-D-galactopyranosyl-

Meloscine **Meloscin**

$C_{19}H_{20}N_2O$
Beilstein · Syst.-Nr. 3572
CAS-Reg.No.: 24314-51-0

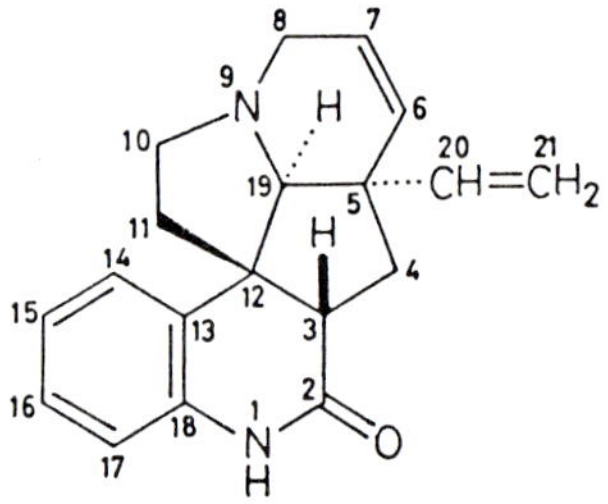

Menthane **Menthan**

sys[CA]: Cyclohexane, methyl(1-methylethyl)-

$C_{10}H_{20}$
Beilstein · Syst.-Nr. 452
CAS-Reg.No.: 52993-54-1
IUPAC: A-72.1

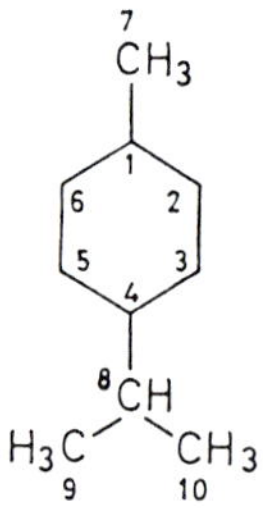

o-Menthane
CAS-Reg.No.: 16580-23-7
cis-o-Menthane
Beilstein: 1-5.150
CAS-Reg.No.: 19489-02-2
trans-o-Menthane
Beilstein: 1-5.150
CAS-Reg.No.: 19489-03-3

m-Menthane
Beilstein: 3-5.132
CAS-Reg.No.: 16580-24-8
cis-m-Menthane
CAS-Reg.No.: 17066-65-8
trans-m-Menthane
CAS-Reg.No.: 17066-66-9

p-Menthane
CAS-Reg.No.: 99-82-1
cis-p-Menthane
Beilstein: 4-5.151
CAS-Reg.No.: 6069-98-3
trans-p-Menthane
Beilstein: 4-5.151
CAS-Reg.No.: 1678-82-6

p-

Menthol **Menthol**

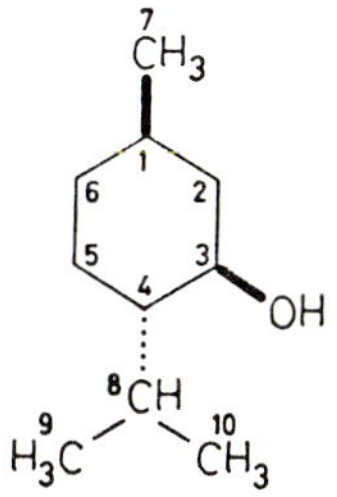

sys[CA]: Cyclohexanol, 5-methyl-2-(1-methylethyl)-, (1α,2β,5α)-

$C_{10}H_{20}O$
Beilstein · Syst.-Nr. 503
CAS-Reg.No.: 89-78-1
IUPAC: C-201.4

(1*R*)-Menthol
Beilstein: 4-6.150
CAS-Reg.No.: 2216-51-5

(1*S*)-Menthol
Beilstein: 4-6.151
CAS-Reg.No.: 15356-60-2

(±)-Menthol
Beilstein: 4-6.151
CAS-Reg.No.: 15356-70-4

(1*R*)-

Mesaconic acid **Mesaconsäure**

= Fumaric acid, methyl-

Mesitylene **Mesitylen**

sys[CA]: Benzene, 1,3,5-trimethyl-

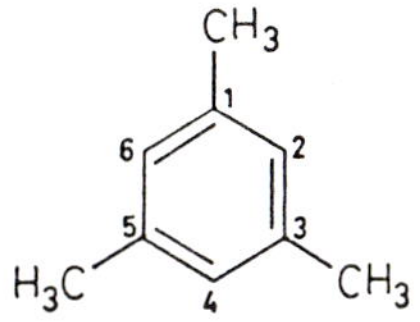

C_9H_{12}
Beilstein: 4-5.1016 · Syst.-Nr. 468
CAS-Reg.No.: 108-67-8
IUPAC: A-12.1

Mesobilirubin **Mesobilirubin**

syn: Mesobilirubin IXα

sys[CA]: 21*H*-Biline-8,12-dipropanoic acid, 2,17-diethyl-1,10,19,-22,23,24-hexahydro-3,7,13,18-tetramethyl-1,19-dioxo-

$C_{33}H_{40}N_4O_6$
Beilstein: 4-26.3261
· Syst.-Nr. 4173
CAS-Reg.No.: 16568-56-2
IUPAC: TP-6.4

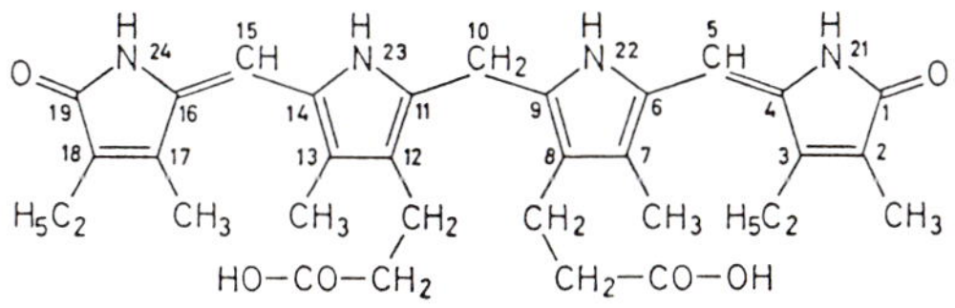

Mesobiliverdin **Mesobiliverdin**

syn: Mesobiliverdin IXα,
Glaucobilin IXα

sys[CA]: 21*H*-Biline-8,12-dipropanoic acid, 3,18-diethyl-1,19,22,24-tetrahydro-2,7,13,17-tetramethyl-1,19-dioxo-

$C_{33}H_{38}N_4O_6$
Beilstein: 4-26.3265
· Syst.-Nr. 4173
CAS-Reg.No.: 29790-16-7
IUPAC: TP-6.4

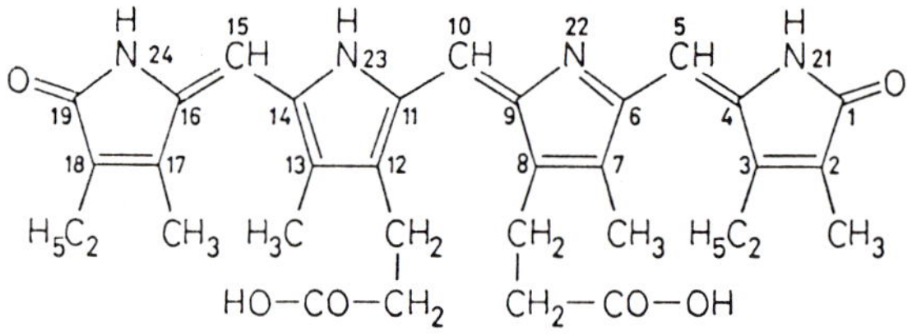

Mesoporphyrin **Mesoporphyrin**

syn: Mesoporphyrin IX

sys[CA]: 21*H*,23*H*-Porphine-2,18-dipropanoic acid, 7,12-diethyl-3,8,13,17-tetramethyl-

$C_{34}H_{38}N_4O_4$
Beilstein: 4-26.3018 · Syst.-Nr. 4173
CAS-Reg.No.: 493-90-3
IUPAC: TP-2.1

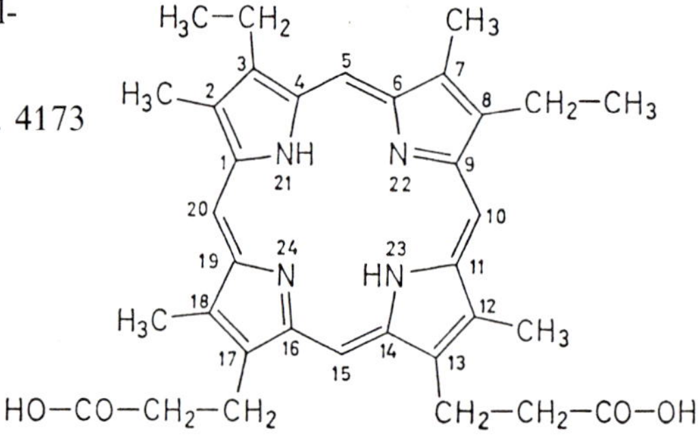

Mesopyropheophorbid a
Mesopyrophäophorbid-a

= Phytochlorin

Mesopyrrochlorin **Mesopyrrochlorin**

= Pyrrochlorin

Mesoxalic acid **Mesoxalsäure**

sys[B]: Malonsäure, Oxo-
[CA]: Propanedioic acid, oxo-

$C_3H_2O_5$
Beilstein: 3-3.1355 · Syst.-Nr. 292
CAS-Reg.No.: 473-90-5
IUPAC: C-416.3

HO – CO – CO – CO – OH

Methacrylic acid **Methacrylsäure**

sys[CA]: 2-Propenoic acid, 2-methyl-

$C_4H_6O_2$ $CH_2=C(CH_3)-CO-OH$

Beilstein: 4-2.1518 · Syst.-Nr. 163

CAS-Reg.No.: 79-41-4

IUPAC: C-404.1

Methane **Methan**

CH_4 CH_4

Beilstein: 4-1.3 · Syst.-Nr. 4

CAS-Reg.No.: 74-82-8

IUPAC: A-1

Methionine **Methionin**

$C_5H_{11}NO_2S$

$H_3C-S-CH_2-CH_2-CH(NH_2)-CO-OH$

Beilstein: 4-4.3189 · Syst.-Nr. 376

CAS-Reg.No.: 7005-18-7

IUPAC: 3AA-1

D-Methionine

Beilstein: 4-4.3189

CAS-Reg.No.: 348-67-4

L-Methionine

Beilstein: 4-4.3189

CAS-Reg.No.: 63-68-3

DL-Methionine

Beilstein: 4-4.3190

CAS-Reg.No.: 59-51-8

Milchsäure = Lactic acid

Morphinan **Morphinan**

$C_{16}H_{21}N$

Beilstein: 4-20.3625 · Syst.-Nr. 3082

CAS-Reg.No.: 468-10-0

IUPAC: F-(46)

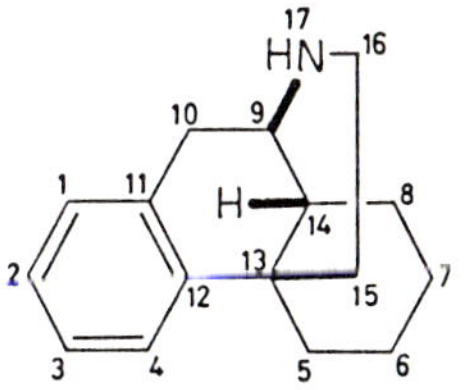

17 HN 16 10 9 1 11 H 14 8 2 13 12 15 7 3 4 5 6

Morpholine **Morpholin**

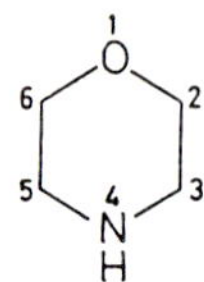

C_4H_9NO
Beilstein: 4-27.15 · Syst.-Nr. 4190 B
CAS-Reg.No.: 110-91-8
IUPAC: B-2.12(14)

Muramic acid **Muramsäure**

$C_9H_{17}NO_7$
Beilstein: 4-4.2029 · Syst.-Nr. 360
CAS-Reg.No.: 1114-41-6
IUPAC: Carb-14

Mycarose **Mycarose**

= *ribo*-2,6-Dideoxy-hexose, 3-methyl-

Myristic acid **Myristinsäure**

= Tetradecanoic acid

Naphthacene **Naphthacen**

$C_{18}H_{12}$
Beilstein: 4-5.2545 · Syst.-Nr. 488
CAS-Reg.No.: 92-24-0
IUPAC: A-21.2(20)

Naphthalene **Naphthalin**

$C_{10}H_8$
Beilstein: 4-5.1640 · Syst.-Nr. 476
CAS-Reg.No.: 91-20-3
IUPAC: A-21.2(3)

Naphthionic acid **Naphthionsäure**

sys[CA]: 1-Naphthalenesulfonic acid, 4-amino-

$C_{10}H_9N_3S$
Beilstein: 4-14.2793 · Syst.-Nr. 1923 F
CAS-Reg.No.: 84-86-6
IUPAC: C-641.1

Naphthoic acid **Naphthoesäure**

sys[CA]: Naphthalenecarboxylic acid

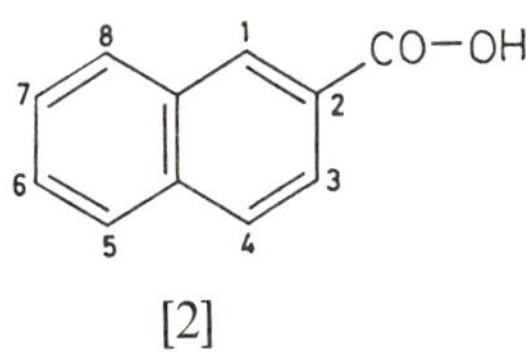

[2]

$C_{11}H_8O_2$
Beilstein · Syst.-Nr. 951
CAS-Reg.No.: 1320-04-3
IUPAC: C-404.1

[1]Naphthoic acid
Beilstein: 4-9.2402
CAS-Reg.No.: 86-55-5

[2]Naphthoic acid
Beilstein: 4-9.2414
CAS-Reg.No.: 93-09-4

Naphthol **Naphthol**

sys[CA]: Naphthalenol

$C_{10}H_8O$
CAS-Reg.No.: 1321-67-1
IUPAC: C-202.2

[2]

[1]Naphthol
Beilstein: 4-6.4208 · Syst.-Nr. 537
CAS-Reg.No.: 90-15-3

[2]Naphthol
Beilstein: 4-6.4253 · Syst.-Nr. 538
CAS-Reg.No.: 135-19-3

Naphthyridine **Naphthyridin**

$C_8H_6N_2$
Beilstein · Syst.-Nr. 3480
IUPAC: B-2.11(29)

[1,5]Naphthyridine
Beilstein: 4-23.1235
CAS-Reg.No.: 254-79-5
[1,6]Naphthyridine
Beilstein: 4-23.1236
CAS-Reg.No.: 253-72-5
[1,7]Naphthyridine
Beilstein: 4-23.1237
CAS-Reg.No.: 253-69-0
[1,8]Naphthyridine
Beilstein: 4-23.1237
CAS-Reg.No.: 254-60-4
[2,6]Naphthyridine
CAS-Reg.No.: 253-50-9
[2,7]Naphthyridine
Beilstein: 4-23.1237
CAS-Reg.No.: 253-45-2

[1,8]

Neogenine **Neogenin**

= Spirostan, (25*S*)-

Neopentane **Neopentan**
sys[CA]: Propane, 2,2-dimethyl-

C_5H_{12}
Beilstein: 4-1.333 · Syst.-Nr. 10
CAS-Reg.No.: 463-82-1
IUPAC: A-2.1

$C(CH_3)_4$

Neuraminic acid **Neuraminsäure**

$C_9H_{17}NO_8$
Beilstein: 4-4.3287 · Syst.-Nr. 377
CAS-Reg.No.: 114-04-5
IUPAC: Lip-4.1

Niacin **Niacin**

= Nicotinic acid

Nicotine **Nicotin**

sys[CA]: Pyridine, 3-(1-methyl-2-pyrrolidinyl)-, (*S*)-

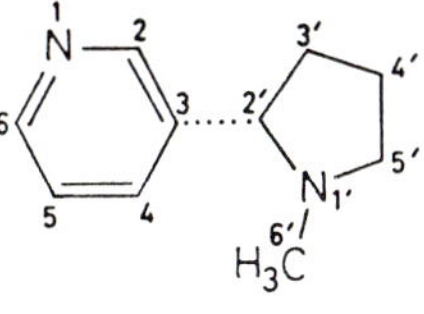

$C_{10}H_{14}N_2$
Beilstein: 4-23.999 · Syst.-Nr. 3470
CAS-Reg.No.: 54-11-5

Nicotinic acid **Nicotinsäure**

sys[CA]: 3-Pyridinecarboxylic acid

$C_6H_5NO_2$
Beilstein: 4-22.348 · Syst.-Nr. 3249
CAS-Reg.No.: 59-67-6
IUPAC: C-404.1; M-8

Nigerose **Nigerose**

= D-Glucose, O^3-α-D-glucopyranosyl-

Nonactin **Nonactin**

$C_{40}H_{64}O_{12}$
Beilstein: 4-19.6279 · Syst.-Nr. 3031
CAS-Reg.No.: 6833-84-7

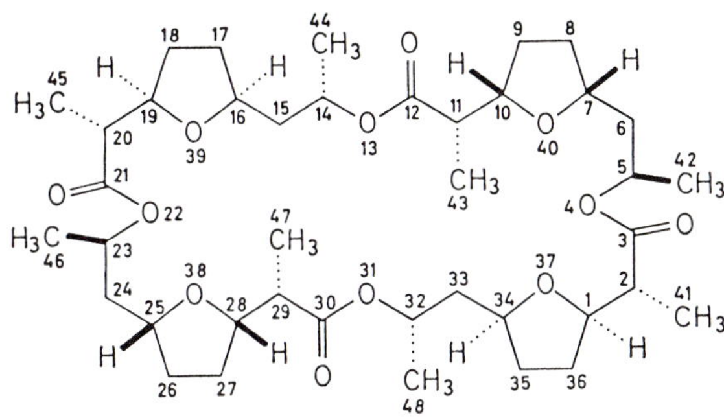

Noraporphine **Noraporphin**

= Aporphane

Norbornane **Norbornan**

sys[CA]: Bicyclo[2.2.1]heptane

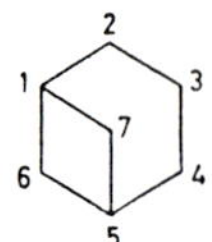

C_7H_{12}
Beilstein: 4-5.258 · Syst.-Nr. 453
CAS-Reg.No.: 279-23-2
IUPAC: A-72.1; according to rule F-4.2 [note] this name should no longer be used (new designation: 8,9,10-trinor-bornan)

Norcarane **Norcaran**

sys[CA]: Bicyclo[4.1.0]heptane

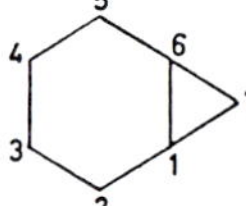

C_7H_{12}
Beilstein: 4-5.257 · Syst.-Nr. 453
CAS-Reg.No.: 286-08-8
IUPAC: A-72.1; according to rule F-4.2 [note] this name should no longer be used (new designation: 8,9,10-trinor-caran)

Norleucine **Norleucin**

sys[B]: Hexansäure, 2-Amino-

$C_6H_{13}NO_2$

$CH_3-[CH_2]_3-CH(NH_2)-CO-OH$

Beilstein · Syst.-Nr. 368
CAS-Reg.No.: 5157-09-5
IUPAC: C-421.1; according to rule 3AA-2.4 no longer recommended

–, 6-oxo- = *Allysine*

D-Norleucine
Beilstein: 4-4.2686
CAS-Reg.No.: 327-56-0

L-Norleucine
Beilstein: 4-4.2686
CAS-Reg.No.: 327-57-1

DL-Norleucine
Beilstein: 4-4.2686
CAS-Reg.No.: 616-06-8

Norpinane **Norpinan**

sys[CA]: Bicyclo[3.1.1]heptane

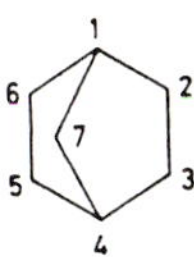

C_7H_{12}
Beilstein · Syst.-Nr. 453
CAS-Reg.No.: 286-34-0
IUPAC: A.72.1; according to rule F-4.2 [note] this name should no longer be used (new designation: 8,9,10-trinor-pinan)

Norvaline **Norvalin**

sys[B]: Valeriansäure, 2-Amino-

$C_5H_{11}NO_2$

$CH_3-CH_2-CH_2-CH(NH_2)-CO-OH$

Beilstein · Syst.-Nr. 367
CAS-Reg.No.: 498-18-0
IUPAC: C-421.1; according to rule 3AA-2.4 no longer recommended

D-Norvaline
Beilstein: 4-4.2629
CAS-Reg.No.: 2013-12-9

L-Norvaline
Beilstein: 4-4.2629
CAS-Reg.No.: 6600-40-4

DL-Norvaline
Beilstein: 4-4.2630
CAS-Reg.No.: 760-78-1

Obacunoic acid **Obacunsäure**

$C_{26}H_{32}O_8$
Beilstein · Syst.-Nr. 2985
CAS-Reg.No.: 751-29-1

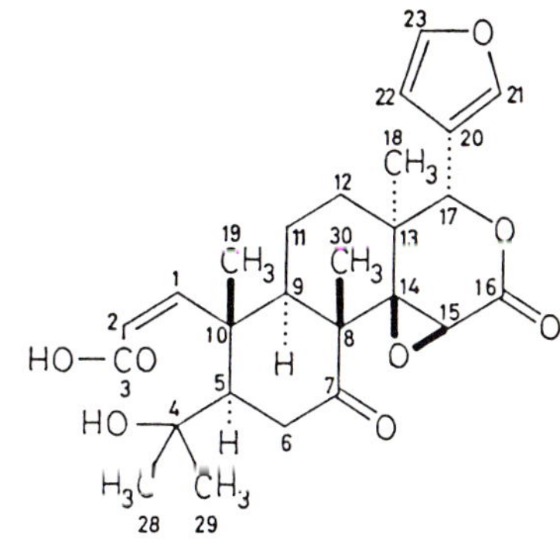

Ochrobirine **Ochrobirin**

$C_{20}H_{19}NO_6$
Beilstein · Syst.-Nr. 4482
CAS-Reg.No.: 24181-64-4

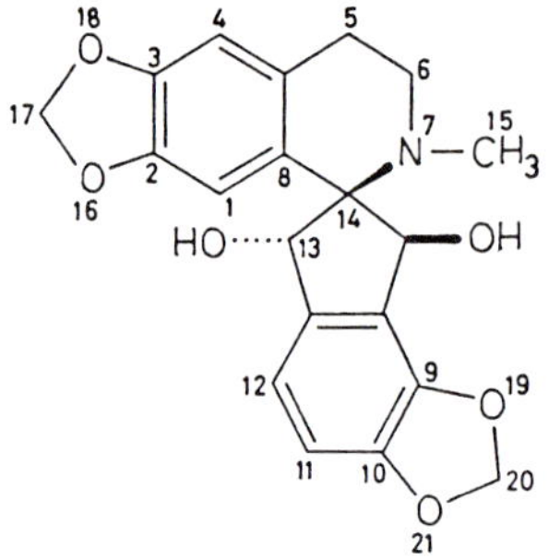

Ölsäure = Oleic acid

Östran = Estrane

Östriol = Estriol

Östron = Estrone

Officinalic acid **Officinalsäure**

$C_{30}H_{44}O_6$
Beilstein · Syst.-Nr. 2896
CAS-Reg.No.: 23983-77-9

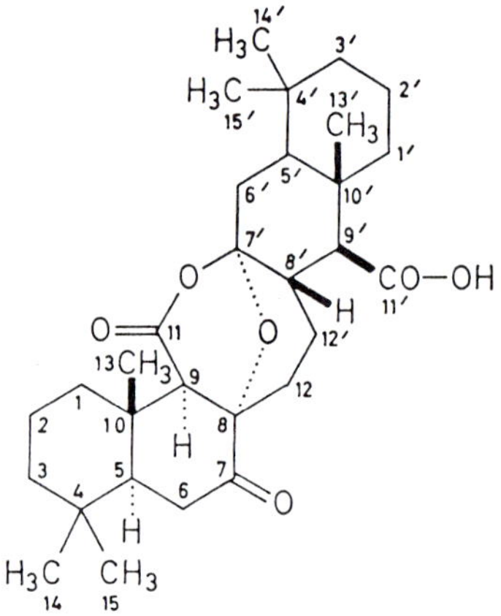

Oleanane **Oleanan**

$C_{30}H_{52}$
Beilstein: 3-5.1341 · Syst.-Nr. 473
CAS-Reg.No.: 471-67-0
IUPAC: F-(20)

–, (14α)-*D*-friedo- = Taraxerane
–, *D:A*-friedo- = Friedelane
–, *D:B*-friedo- = *Glutinane*

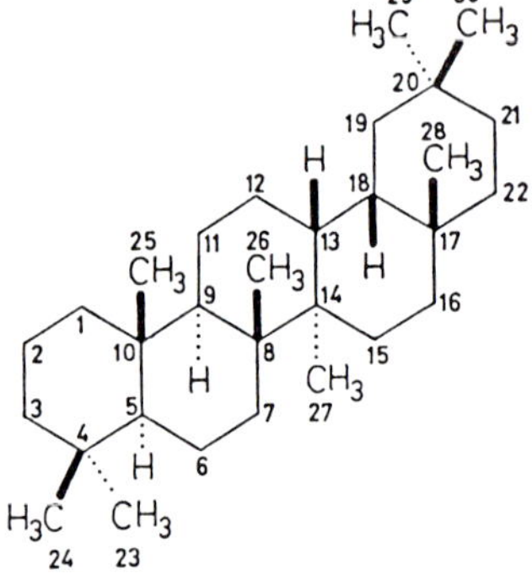

Oleandolide **Oleandolid**

$C_{20}H_{34}O_7$
Beilstein · Syst.-Nr. 2843
CAS-Reg.No.: 68540-16-9

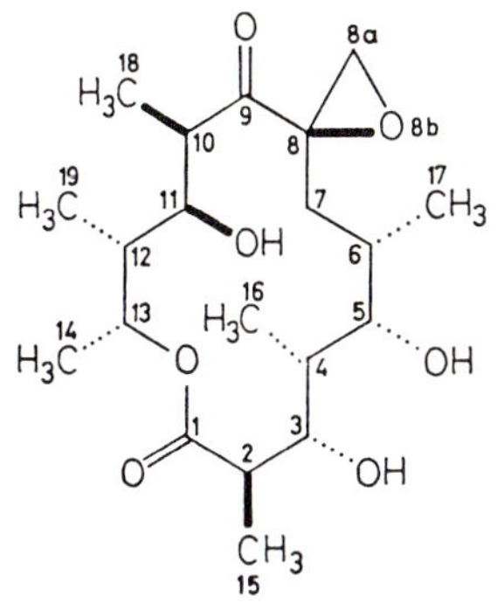

Oleandrose **Oleandrose**

= *arabino*-2,6-Dideoxy-hexose, O^3-methyl-

Oleic acid **Ölsäure**

sys[B]: Octadec-9*c*-ensäure
[CA]: 9-Octadecenoic acid, (*Z*)-

$C_{18}H_{34}O_2$
Beilstein: 4-2.1641 · Syst.-Nr. 163
CAS-Reg.No.: 112-80-1
IUPAC: C-404.1

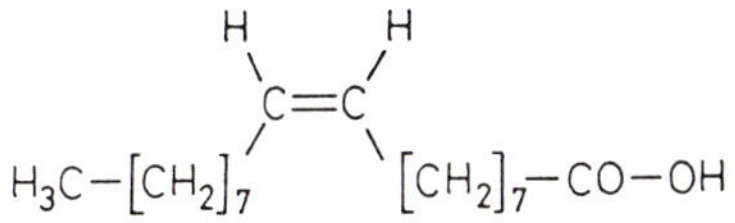

Olivomose **Olivomose**

= *lyxo*-2,6-Dideoxy-hexose, O^4-methyl-

Olivomycose **Olivomycose**

= *arabino*-2,6-Dideoxy-hexose, 3-methyl-

Olivose **Olivose**

= *arabino*-2,6-Dideoxy-hexose

Onocerane **Onoceran**

sys[B+CA]: 8,14-Secogammacerane

$C_{30}H_{54}$
Beilstein: 4-5.1240 · Syst.-Nr. 473
CAS-Reg.No.: 511-03-5

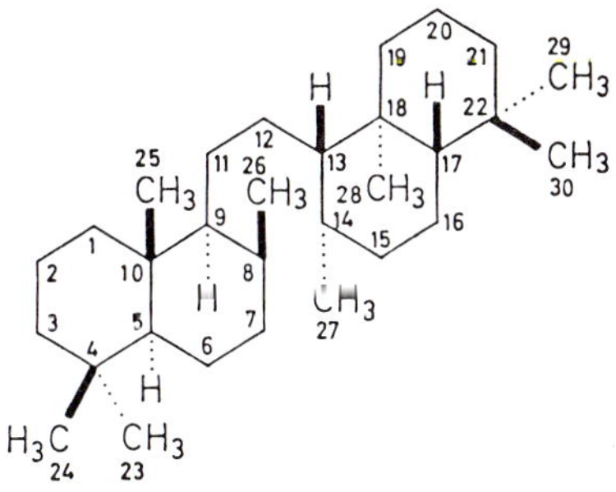

Ophiobolane **Ophiobolan**

$C_{25}H_{46}$
Beilstein: 4-18.1619* · Syst.-Nr. 461
CAS-Reg.No.: 20098-65-1
IUPAC: F-2

Ormosanine **Ormosanin**

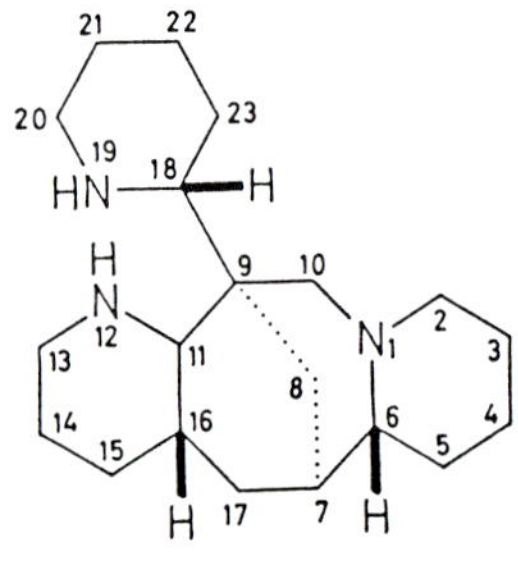

$C_{20}H_{35}N_3$
Beilstein: 4-26.93 · Syst.-Nr. 3800
CAS-Reg.No.: 5001-21-8

Ornithine **Ornithin**
sys[B]: Valeriansäure, 2,5-Diamino-

$C_5H_{12}N_2O_2$

$H_2N-[CH_2]_3-CH(NH_2)-CO-OH$

Beilstein · Syst.-Nr. 367
CAS-Reg.No.: 7006-33-9
IUPAC: C-421.1; 3AA-Appendix

–, N^5-carbamoyl- = *Citrulline*

D-Ornithine
Beilstein: 4-4.2644
CAS-Reg.No.: 348-66-3

L-Ornithine
Beilstein: 4-4.2644
CAS-Reg.No.: 70-26-8

DL-Ornithine
Beilstein: 4-4.2645
CAS-Reg.No.: 616-07-9

Orotic acid **Orotsäure**

sys[B]: Pyrimidin-4-carbonsäure, 2,6-Dioxo-1,2,3,6-tetrahydro-
[CA]: 4-Pyrimidinecarboxylic acid, 1,2,3,6-tetrahydro-2,6-dioxo-

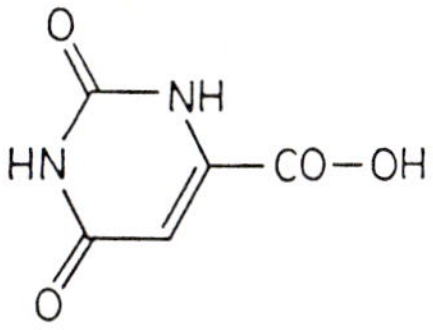

$C_5H_4N_2O_4$
Beilstein: 4-25.1759 · Syst.-Nr. 3697
CAS-Reg.No.: 65-86-1

Orotidine **Orotidin**

sys[B]: Pyrimidin-4-carbonsäure, 2,6-Dioxo-3-β-D-ribofuranosyl-1,2,3,6-tetrahydro-
[CA]: 4-Pyrimidinecarboxylic acid, 1,2,3,6-tetrahydro-2,6-dioxo-3-β-D-ribofuranosyl-

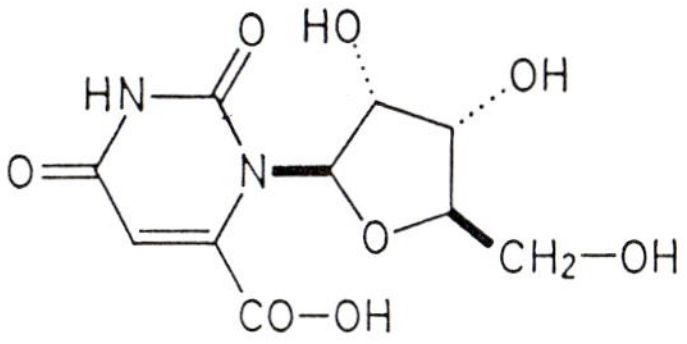

$C_{10}H_{12}N_2O_8$
Beilstein: 4-25.1763 · Syst.-Nr. 3697
CAS-Reg.No.: 314-50-1

Osmane **Osman**

= Iridane

Ovalene **Ovalen**

$C_{32}H_{14}$
Beilstein: 4-5.2944 · Syst.-Nr. 497
CAS-Reg.No.: 190-26-1
IUPAC: A-21.2(35)

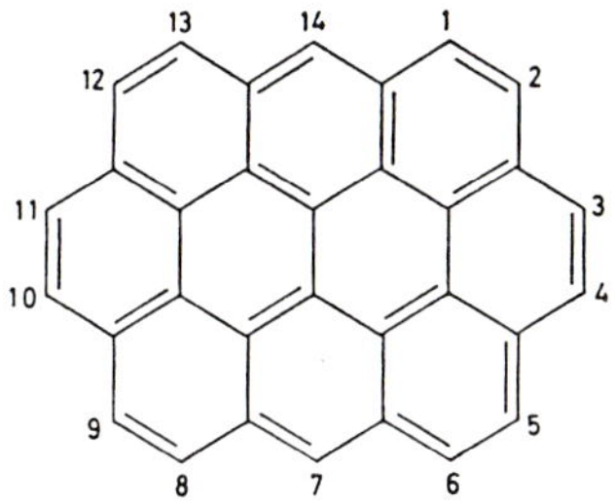

Oxalacetic acid **Oxalessigsäure**

sys[B]: Bernsteinsäure, Oxo-
[CA]: Butanedioic acid, oxo-

$C_4H_4O_5$

$HO-CO-CH_2-CO-CO-OH$

Beilstein: 4-3.1808 · Syst.-Nr. 292
CAS-Reg.No.: 328-42-7
IUPAC: C-416.3

Oxalamic acid **Oxalamidsäure**

= Oxamic acid

Oxalic acid **Oxalsäure**

sys[CA]: Ethanedioic acid

$C_2H_2O_4$ HO – CO – CO – OH

Beilstein: 4-2.1819 · Syst.-Nr. 170

CAS-Reg.No.: 144-62-7

IUPAC: C-404.1

Oxamic acid **Oxamidsäure**

sys[B]: Oxalamidsäure

[CA]: Acetic acid, aminooxo-

$C_2H_3NO_3$ H_2N – CO – CO – OH

Beilstein: 4-2.1857 · Syst.-Nr. 170

CAS-Reg.No.: 471-47-6

IUPAC: C-431.1

Oxyacanthan **Oxyacanthan**

$C_{32}H_{30}N_2O_2$

Beilstein: 4-19.4280* · Syst.-Nr. 4635

CAS-Reg.No.: 34479-48-6

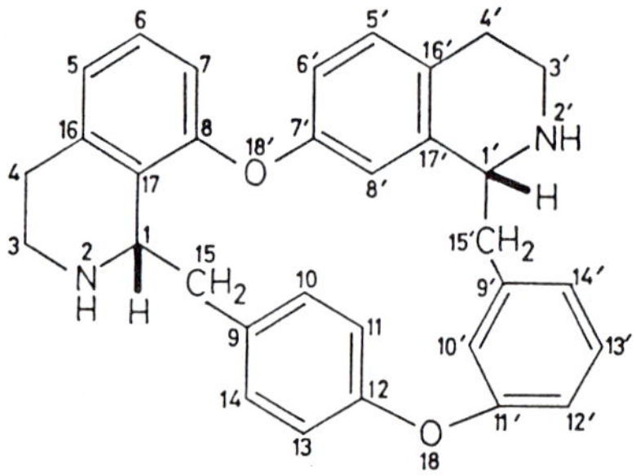

Palmitic acid **Palmitinsäure**

= Hexadecanoic acid

Panamine **Panamin**

$C_{20}H_{33}N_3$

Beilstein: 4-26.165 · Syst.-Nr. 3808

CAS-Reg.No.: 2448-27-3

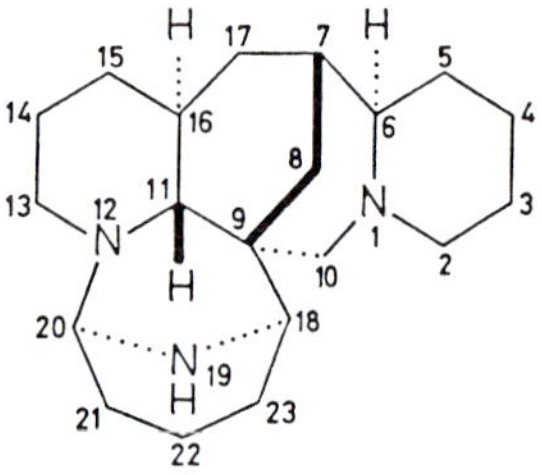

Pancracine **Pancracin**

$C_{16}H_{17}NO_4$
Beilstein: 4-27.6461 · Syst.-Nr. 4441
CAS-Reg.No.: 21416-14-8

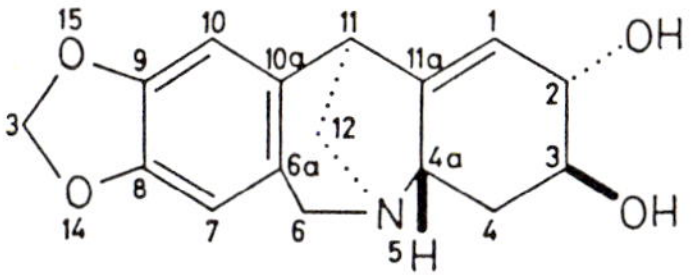

Paratose **Paratose**

= D-*ribo*-3,6-Dideoxy-hexose

Parsonsine **Parsonsin**

$C_{22}H_{33}NO_8$
Beilstein · Syst.-Nr. 4475
CAS-Reg.No.: 72213-98-0

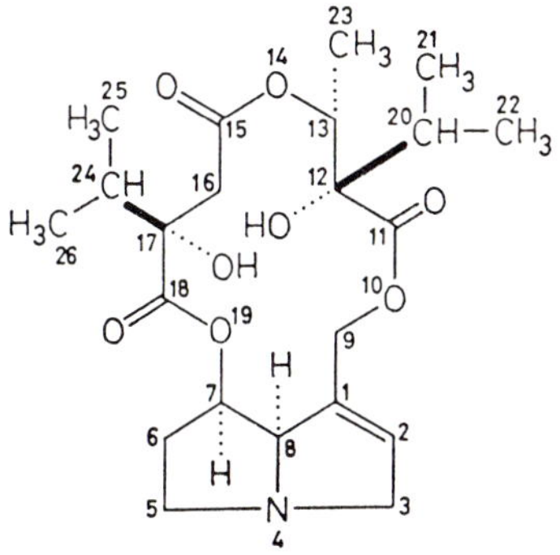

Penicillanic acid **Penicillansäure**

sys[CA]: 4-Thia-1-azabicyclo[3.2.0]-heptane-2-carboxylic acid, 3,3-dimethyl-7-oxo-, (2*S*-*cis*)-

$C_8H_{11}NO_3S$
Beilstein: 4-27.5858* · Syst.-Nr. 4330
CAS-Reg.No.: 87-53-6

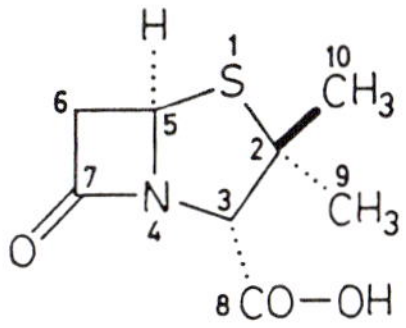

Pentaerythritol **Pentaerythrit**

sys[CA]: 1,3-Propanediol, 2,2-bis(hydroxymethyl)-

$C_5H_{12}O_4$
Beilstein: 4-1.2812 · Syst.-Nr. 47
CAS-Reg.No.: 115-77-5
IUPAC: C-201.4

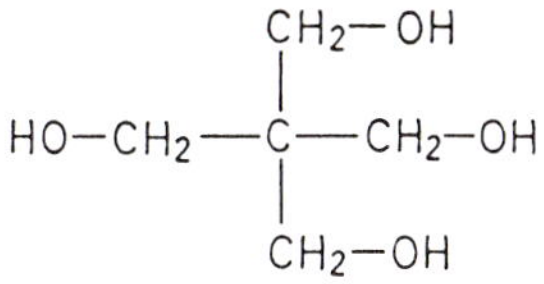

Pentalene **Pentalen**

C_8H_6
Beilstein · Syst.-Nr. 474
CAS-Reg.No.: 250-25-9
IUPAC: A-21.2(1)

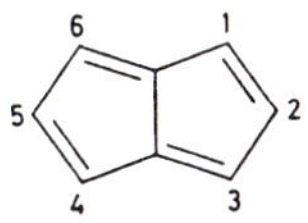

Pentarane **Pentaran**

= Benzo[16,17]androstane, hexahydro-

Perimidine **Perimidin**

$C_{11}H_8N_2$
Beilstein: 4-23.1570 · Syst.-Nr. 3486
CAS-Reg.No.: 204-02-4
IUPAC: B-2.11(39)

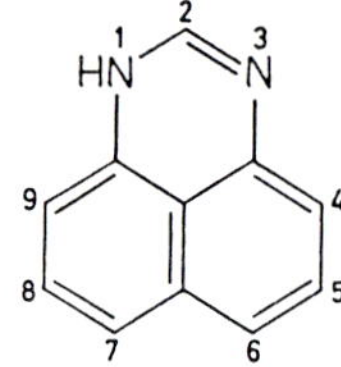

Perylene **Perylen**

$C_{20}H_{12}$
Beilstein: 4-5.2689 · Syst.-Nr. 490
CAS-Reg.No.: 198-55-0
IUPAC: A-21.2(23)

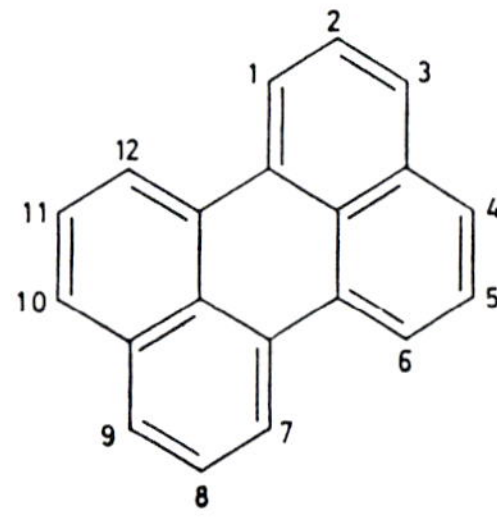

Phäophorbid = Pheophorbide

Phenethyl alcohol **Phenäthylalkohol**
sys[CA]: Benzeneethanol

$C_8H_{10}O$
Beilstein: 4-6.3067 · Syst.-Nr. 529
CAS-Reg.No.: 60-12-8
IUPAC: C-201.4

CH_2-CH_2-OH

Phenalene **Phenalen**

$C_{13}H_{10}$
Beilstein: 3-5.1953 · Syst.-Nr. 480
CAS-Reg.No.: 203-80-5
IUPAC: A-21.2(11)

Phenanthrazine **Phenanthrazin**
sys[B]: Tetrabenzo[*a,c,h,i*]phenazin

$C_{28}H_{16}N_2$
Beilstein: 4-23.2203 · Syst.-Nr. 3499
CAS-Reg.No.: 215-14-5

Phenanthrene **Phenanthren**

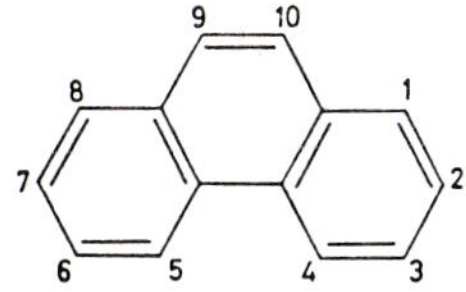

$C_{14}H_{10}$
Beilstein: 4-5.2297 · Syst.-Nr. 485
CAS-Reg.No.: 85-01-8
IUPAC: A-21.2(12)

Phenanthridine **Phenanthridin**

$C_{13}H_9N$, formula A [former numbering: formula B]
Beilstein: 4-20.4016 · Syst.-Nr. 3088
CAS-Reg.No.: 229-87-8
IUPAC: B-2.11(37)

A

B

Phenanthroline **Phenanthrolin**

$C_{12}H_8N_2$
Beilstein · Syst.-Nr. 3487
IUPAC: B-2.11(40)

[1,5]Phenanthroline = Benzo[*c*][1,5]naphthyridine
[1,6]Phenanthroline = Benzo[*h*][1,6]naphthyridine

[1,7]Phenanthroline
Beilstein: 4-23.1675
CAS-Reg.No.: 230-46-6

[1,7]

[1,8]Phenanthroline
Beilstein: 4-23.1677
CAS-Reg.No.: 230-30-8

[1,9]Phenanthroline
CAS-Reg.No.: 230-45-5

[1,10]Phenanthroline
Beilstein: 4-23.1677
CAS-Reg.No.: 66-71-7

[2,5]Phenanthroline =
Benzo[*c*][1,6]naphthyridine
[2,6]Phenanthroline =
Benzo[*c*][2,6]naphthyridine

[2,7]Phenanthroline
CAS-Reg.No.: 10284-69-2

[2,8]Phenanthroline
CAS-Reg.No.: 668-87-1

[2,9]Phenanthroline
CAS-Reg.No.: 19376-27-3

[3,5]Phenanthroline =
Benzo[*c*][1,7]naphthyridine
[3,6]Phenanthroline =
Benzo[*c*][2,7]naphthyridine

[3,7]Phenanthroline
CAS-Reg.No.: 19376-21-7

[3,8]Phenanthroline
Beilstein: 4-23.1683
CAS-Reg.No.: 229-70-9

[4,5]Phenanthroline =
Benzo[*c*][1,8]naphthyridine
[4,6]Phenanthroline =
Benzo[*f*][1,7]naphthyridine

[4,7]Phenanthroline
Beilstein: 4-23.1683
CAS-Reg.No.: 230-07-9

[5,6]Phenanthroline =
Benzo[*c*]cinnoline

Phenarsazine **Phenarsazin**

$C_{12}H_8AsN$, formula A [former numbering: formula B]
Beilstein: 4-27.9797 · Syst.-Nr. 4720
CAS-Reg.No.: 260-24-2
IUPAC: B-2.11(42)

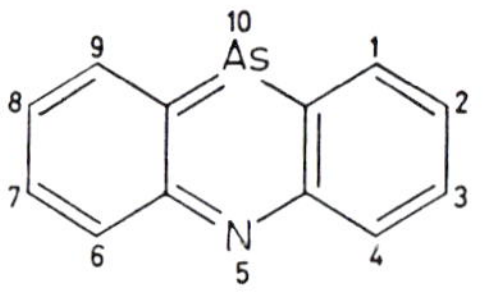

A

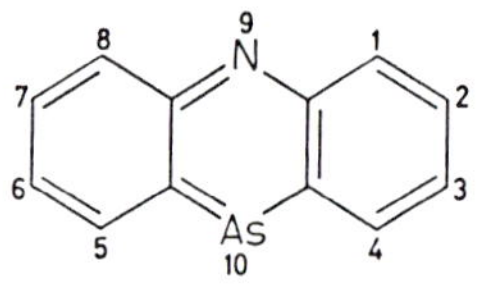

B

Phenazine **Phenazin**

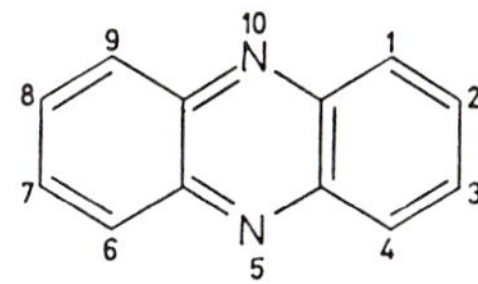

$C_{12}H_8N_2$
Beilstein: 4-23.1654 · Syst.-Nr. 3487
CAS-Reg.No.: 92-82-0
IUPAC: B-2.11(41)

Phenetidine **Phenetidin**
sys[CA]: Benzenamine, *ar*-ethoxy-

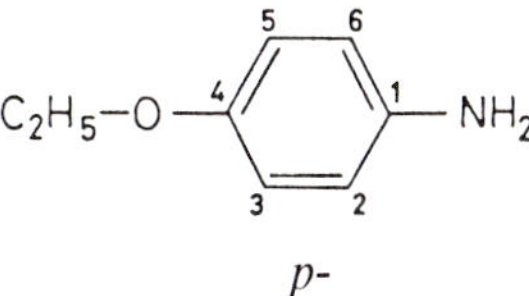

$C_8H_{11}NO$
CAS-Reg.No.: 1321-31-9
IUPAC: C-812.1

o-Phenetidine
Beilstein: 4-13.807 · Syst.-Nr. 1829
CAS-Reg.No.: 94-70-2
m-Phenetidine
Beilstein: 4-13.954 · Syst.-Nr. 1840
CAS-Reg.No.: 621-33-0
p-Phenetidine
Beilstein: 4-13.1017 · Syst.-Nr. 1843
CAS-Reg.No.: 156-43-4

Phenetole **Phenetol**
sys[CA]: Benzene, ethoxy-

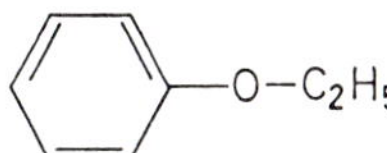

$C_8H_{10}O$
Beilstein: 4-6.554 · Syst.-Nr. 514
CAS-Reg.No.: 103-73-1
IUPAC: C-214.1

Phenol **Phenol**

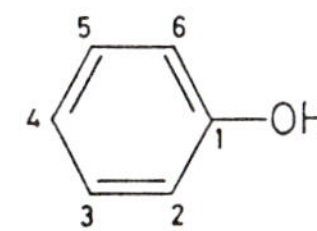

C_6H_6O
Beilstein: 4-6.531 · Syst.-Nr. 512
CAS-Reg.No.: 108-95-2
IUPAC: C-202.2

Phenothiazine **Phenothiazin**
sys[CA]: 10*H*-Phenothiazine

$C_{12}H_9NS$
Beilstein: 4-27.1214 · Syst.-Nr. 4198
CAS-Reg.No.: 92-84-2
IUPAC: B-2.11(44)

1*H*-Phenothiazine
CAS-Reg.No.: 261-86-9

Phenoxathiin **Phenoxathiin**

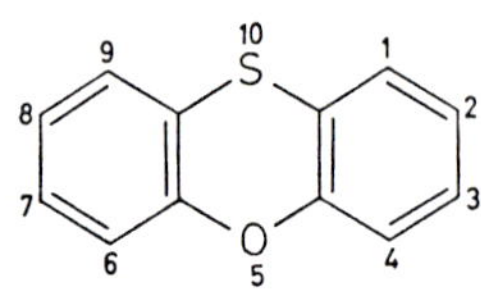

$C_{12}H_8OS$
Beilstein: 4-19.341 · Syst.-Nr. 2676
CAS-Reg.No.: 262-20-4
IUPAC: B-2.11(10)

Phenoxazine **Phenoxazin**

sys[CA]: 10*H*-Phenoxazine

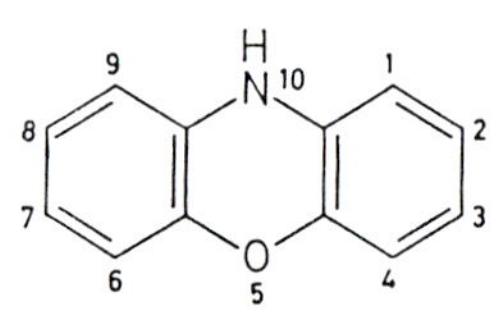

$C_{12}H_9NO$
Beilstein: 4-27.1209 · Syst.-Nr. 4198
CAS-Reg.No.: 135-67-1
IUPAC: B-2.11(47)

1*H*-Phenoxazine
CAS-Reg.No.: 261-76-7
3*H*-Phenoxazine
CAS-Reg.No.: 261-77-8

Phenylalanine **Phenylalanin**

$C_9H_{11}NO_2$
Beilstein · Syst.-Nr. 1905 G
CAS-Reg.No.: 3617-44-5
IUPAC: 3AA-2.2.3

β α
$-CH_2-CH(NH_2)-CO-OH$

D-Phenylalanine
Beilstein: 4-14.1552
CAS-Reg.No.: 673-06-3
L-Phenylalanine
Beilstein: 4-14.1552
CAS-Reg.No.: 63-91-2
DL-Phenylalanine
Beilstein: 4-14.1553
CAS-Reg.No.: 150-30-1

Pheophorbide a **Phäophorbid-a**

sys[CA]: 3-Phorbinepropanoic acid, 9-ethenyl-14-ethyl-21-(methoxycarbonyl)-4,8,13,18-tetramethyl-20-oxo-, [3*S*-(3α,4β,21β)]-

$C_{35}H_{36}N_4O_5$
Beilstein: 4-26.3237 · Syst.-Nr. 4173
CAS-Reg.No.: 15664-29-6
IUPAC: TP-4.3.2

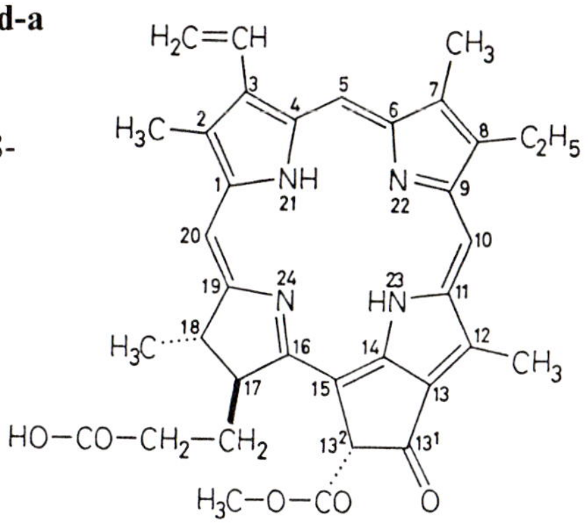

Pheophorbide b **Phäophorbid-b**

sys[CA]: 3-Phorbinepropanoic acid, 9-ethenyl-14-ethyl-13-formyl-21-(methoxycarbonyl)-4,8,18-trimethyl-20-oxo-, [3*S*-(3α,4β,21β)]-

$C_{35}H_{34}N_4O_6$
Beilstein: 4-26.3284 · Syst.-Nr. 4173
CAS-Reg.No.: 20239-99-0
IUPAC: TP-4.3.2

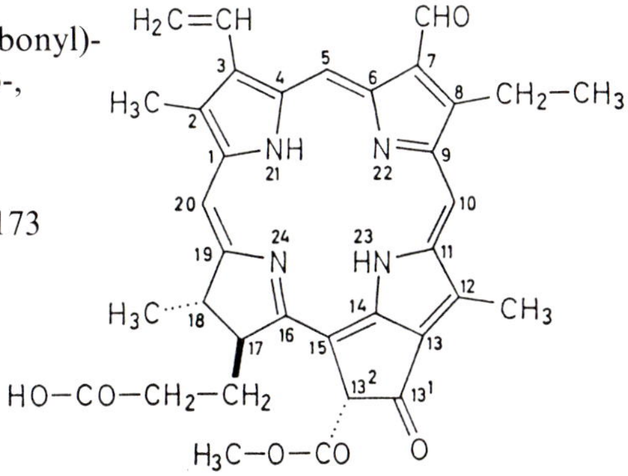

Phlorin **Phlorin**

= Porphyrin, 5,22-dihydro-

Phloroglucinol **Phloroglucin**

sys[CA]: 1,3,5-Benzenetriol

$C_6H_6O_3$
Beilstein: 4-6.7361 · Syst.-Nr. 580
CAS-Reg.No.: 108-73-6
IUPAC: C-202.2

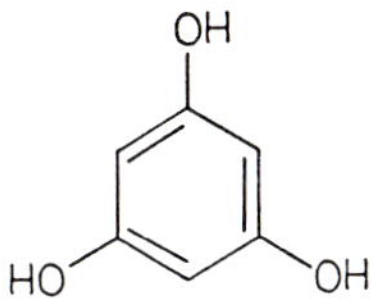

Phorbine **Phorbin**

$C_{22}H_{18}N_4$
Beilstein · Syst.-Nr. 4031
CAS-Reg.No.: 24861-47-0
IUPAC: TP-1.3

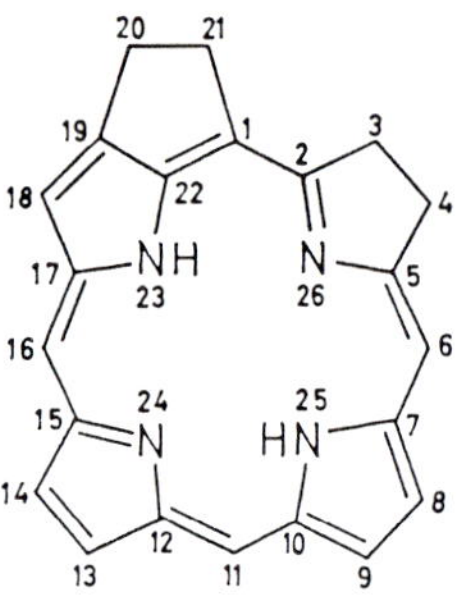

Phthalan **Phthalan**

sys[CA]: Isobenzofuran, 1,3-dihydro-

C_8H_8O
Beilstein: 5-17/1.584 · Syst.-Nr. 2366
CAS-Reg.No.: 496-14-0

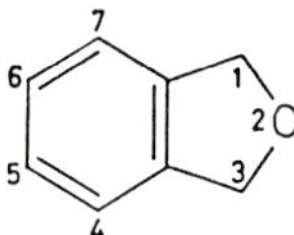

Phthalazine **Phthalazin**

$C_8H_6N_2$
Beilstein: 4-23.1233 · Syst.-Nr. 3480
CAS-Reg.No.: 253-52-1
IUPAC: B-2.11(28)

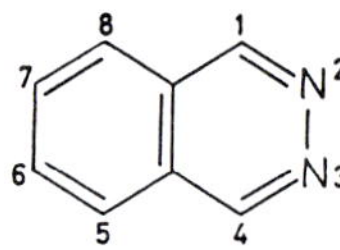

Phthalic acid **Phthalsäure**

sys[CA]: 1,2-Benzenedicarboxylic acid

$C_8H_6O_4$
Beilstein: 4-9.3167 · Syst.-Nr. 970
CAS-Reg.No.: 88-99-3
IUPAC: C-404.1

Phthalide **Phthalid**

sys[B]: Phthalan-1-on
[CA]: 1(3*H*)-Isobenzofuranone

$C_8H_6O_2$
Beilstein: 4-17.4948 · Syst.-Nr. 2463
CAS-Reg.No.: 87-41-2
IUPAC: C-473.1

Phthalocyanine **Phthalocyanin**

$C_{32}H_{18}N_8$
Beilstein: 4-26.4255 · Syst.-Nr. 4187
CAS-Reg.No.: 574-93-6
IUPAC: TP-1.6

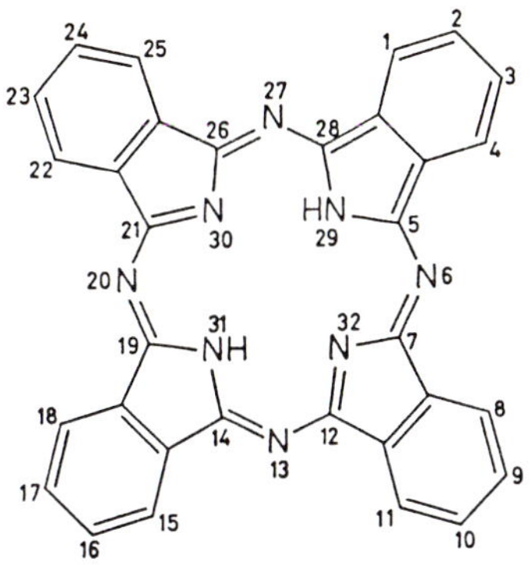

Phthaloperine **Phthaloperin**

$C_{18}H_{12}N_2$
Beilstein · Syst.-Nr. 3491
CAS-Reg.No.: 42315-76-4

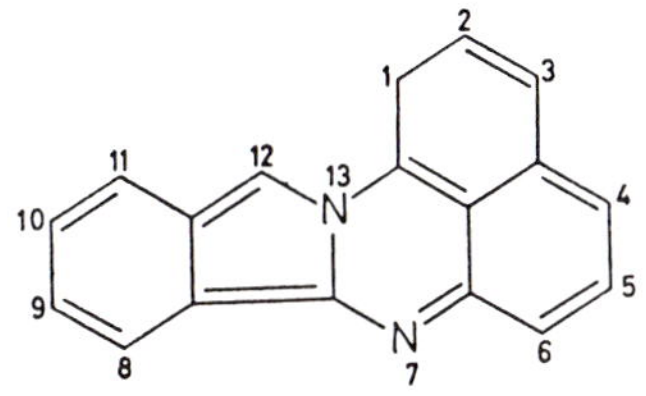

Phthalsäure = Phthalic acid

Phyllochlorin **Phyllochlorin**

sys[CA]: 21*H*,23*H*-Porphine-2-propanoic acid, 8,13-diethyl-2,3-dihydro-3,7,12,17,20-pentamethyl-, (2*S*-*trans*-)

$C_{32}H_{38}N_4O_2$
Beilstein: 4-26.2943 · Syst.-Nr. 4173
IUPAC: TP-4.3.1

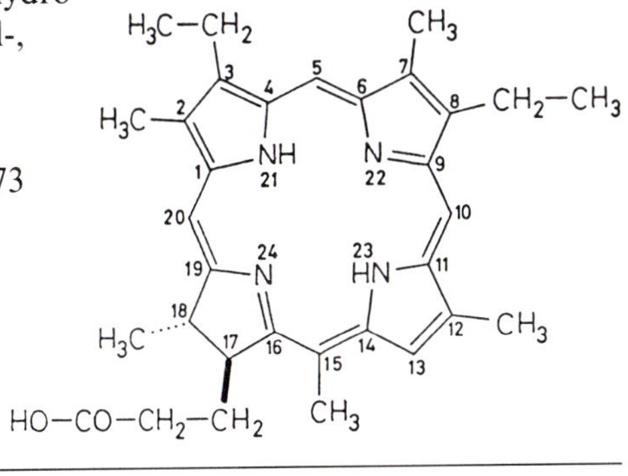

Phyllocladane **Phyllocladan**

sys[CA]: Kaurane, (5α,9α,10β)-

$C_{20}H_{34}$
Beilstein: 4-5.1214 · Syst.-Nr. 472
CAS-Reg.No.: 469-84-1
IUPAC: F-2

–, 13-methyl-17-nor- = *Beyerane, Stachane*

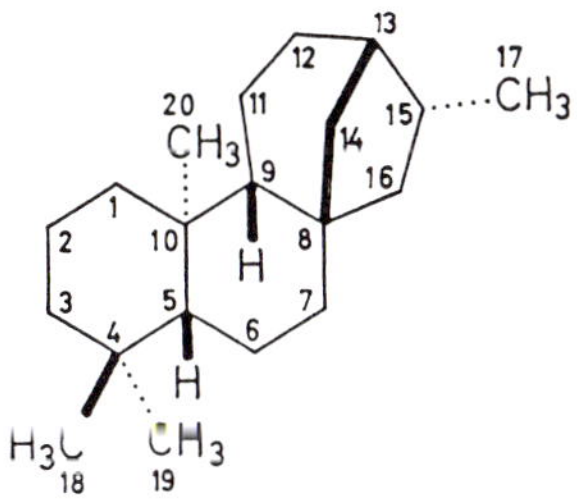

Phylloerythrin **Phylloerythrin**

= Phytoporphyrin

Phylloporphyrin **Phylloporphyrin**

syn: γ-Phylloporphyrin XV
sys[CA]: 21*H*,23*H*-Porphine-2-propanoic acid, 8,13-diethyl-3,7,12,17,20-pentamethyl-

$C_{32}H_{36}N_4O_2$
Beilstein: 4-26.2959 · Syst.-Nr. 4173
CAS-Reg.No.: 13939-70-3
IUPAC: TP-2.1

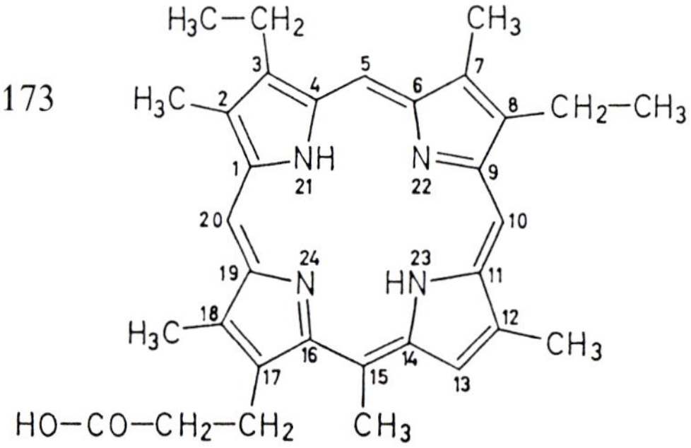

Phytochlorin **Phytochlorin**

syn: Mesopyropheophorbide a
sys[CA]: 3-Phorbinepropanoic acid, 9,14-diethyl-4,8,13,18-tetramethyl-20-oxo-, (3*S*-*trans*)-

$C_{33}H_{36}N_4O_3$
Beilstein: 4-26.3184 · Syst.-Nr. 4173
CAS-Reg.No.: 56145-41-6
IUPAC: TP-4.3.1

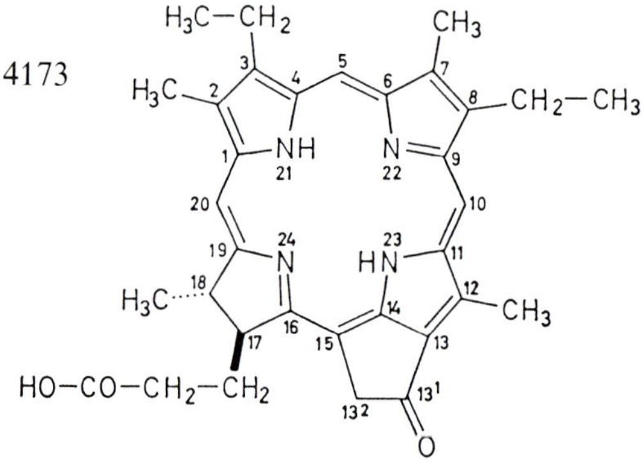

Phytol **Phytol**

sys[CA]: 2-Hexadecen-1-ol, 3,7,11,15-tetramethyl-, [*R*-[*R**,*R**-(*E*)]]

$C_{20}H_{40}O$
Beilstein: 4-1.2208 · Syst.-Nr. 25
CAS-Reg.No.: 150-86-7
IUPAC: C-201.4

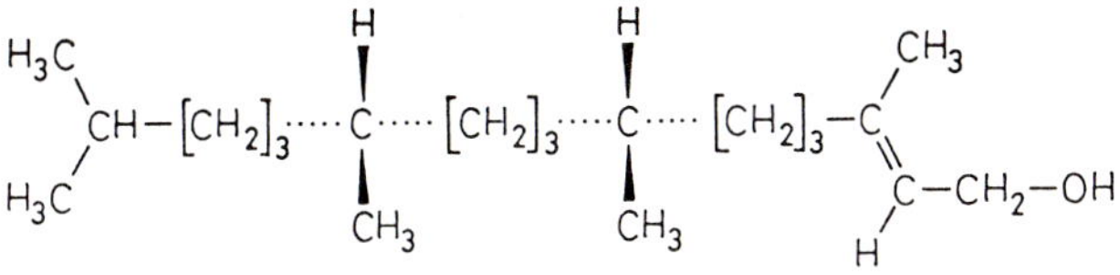

Phytoporphyrin **Phytoporphyrin**

syn: Phylloerythrin
sys[CA]: 3-Phorbinepropanoic acid, 3,4-didehydro-9,14-diethyl-4,8,13,18-tetramethyl-20-oxo-

$C_{33}H_{34}N_4O_3$
Beilstein: 4-26.3188 · Syst.-Nr. 4173
CAS-Reg.No.: 26359-43-3
IUPAC: TP-2.1

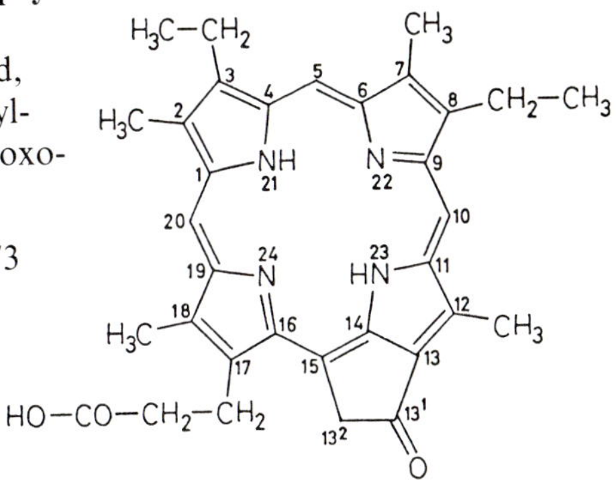

Picene **Picen**

$C_{22}H_{14}$
Beilstein: 4-5.2724 · Syst.-Nr. 491
CAS-Reg.No.: 213-46-7
IUPAC: A-21.2(22)

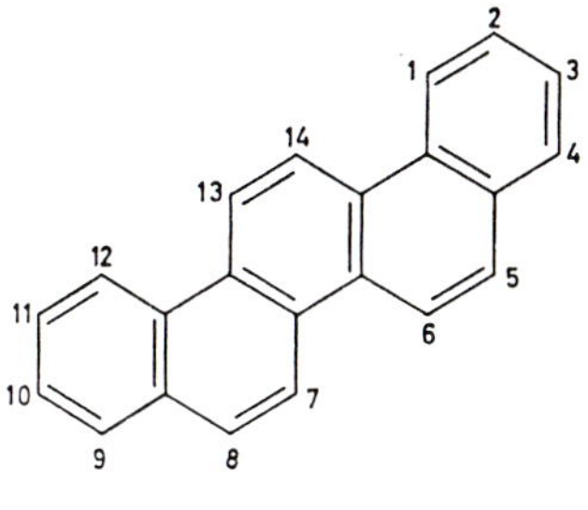

Picrasane **Picrasan**

$C_{20}H_{34}O$
Beilstein: 4-18.2531* · Syst.-Nr. 2365
CAS-Reg.No.: 35732-97-9
IUPAC: F-2

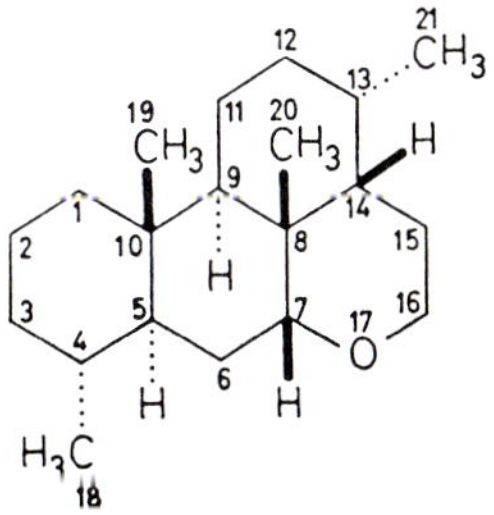

Picric acid **Picrinsäure**

sys[CA]: Phenol, 2,4,6-trinitro-

$C_6H_4N_3O_7$
Beilstein: 4-6.1388 · Syst.-Nr. 523
CAS-Reg.No.: 88-89-1
IUPAC: C-202.2

Pimarane **Pimaran**

sys[CA]: Phenanthrene, 7-ethyl-tetradecahydro-1,1,4a,7-tetramethyl-, [4a*R*-(4aα,4bβ,7α,-8aα,10aβ)]-

$C_{20}H_{36}$, formula A [former numbering: formula B (configuration at C-13 not defined)]
Beilstein: 5-17/1.677* · Syst.-Nr. 461
CAS-Reg.No.: 30257-03-5
IUPAC: F-2

–, 9-methyl-20-nor-(8α,9β)- = Rosane
–, 14-methyl-17-nor- = *Cassane*

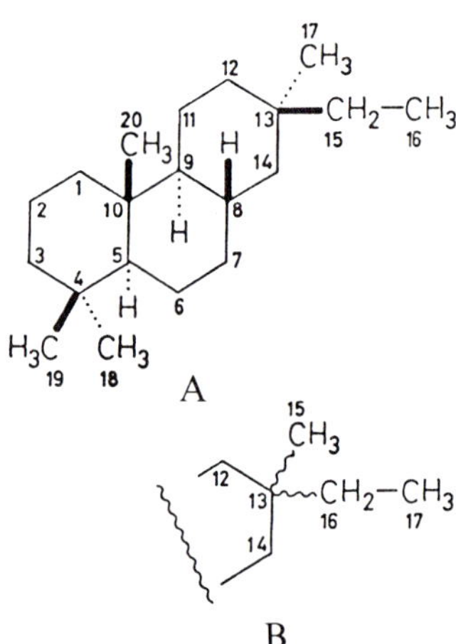

Pimelic acid **Pimelinsäure**

= Heptanedioic acid

Pinane **Pinan**

sys[CA]: Bicyclo[3.1.1]heptane, 2,6,6-trimethyl-

$C_{10}H_{18}$
Beilstein: 4-5.318 · Syst.-Nr. 453
CAS-Reg.No.: 473-55-2
IUPAC: A-72.1

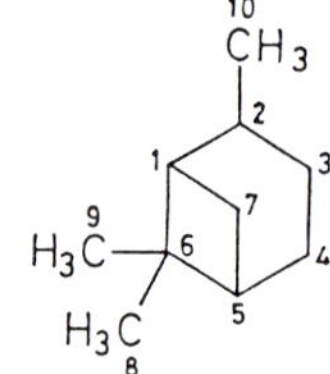

Piperazine **Piperazin**

$C_4H_{10}N_2$
Beilstein: 4-23.15 · Syst.-Nr. 3460 B
CAS-Reg.No.: 110-85-0
IUPAC: B-2.12(10)

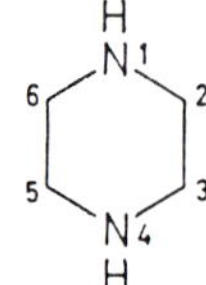

Piperidine **Piperidin**

$C_5H_{11}N$
Beilstein: 4-20.287 · Syst.-Nr. 3038
CAS-Reg.No.: 110-89-4
IUPAC: B-2.12(9)

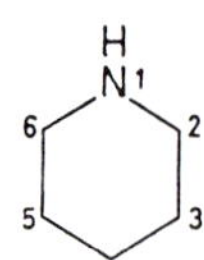

Piperinsäure = Piperonic acid

Piperonal **Piperonal**

sys[CA]: 1,3-Benzodioxole-5-carbox-aldehyde

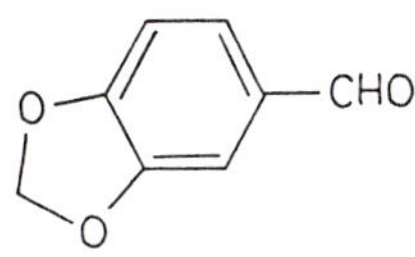

$C_8H_6O_3$
Beilstein: 4-19.1649 · Syst.-Nr. 2742
CAS-Reg.No.: 120-57-0
IUPAC: C-305.2

Piperonic acid **Piperinsäure**

sys[B]: Penta-2,4-diensäure, 5-Benzo[1,3]dioxol-5-yl-
[CA]: 2,4-Pentadienoic acid, 5-(1,3-benzodioxol-5-yl)-

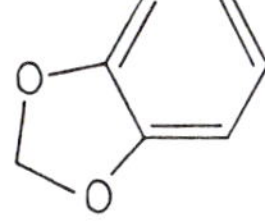

$C_{12}H_{10}O_4$
Beilstein: 4-19.3565 · Syst.-Nr. 2852
CAS-Reg.No.: 5285-18-7

Piperonylic acid **Piperonylsäure**

sys[CA]: 1,3-Benzodioxole-5-carboxylic acid

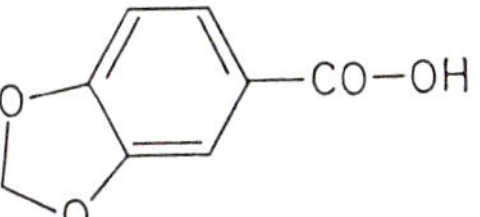

$C_8H_6O_4$
Beilstein: 4-19.3493 · Syst.-Nr. 2850
CAS-Reg.No.: 94-53-1
IUPAC: C-411.1

Pivalic acid **Pivalinsäure**

= Propionic acid, 2,2-dimethyl-

Pleiadene **Pleiaden**

$C_{18}H_{12}$
Beilstein · Syst.-Nr. 488
CAS-Reg.No.: 206-92-8
IUPAC: A-21.2(21)

Podocarpane **Podocarpan**

sys[CA]: Phenanthrene, tetradeca-hydro-1,1,4a-trimethyl-, [4a*R*-(4aα,4bβ,8aα,10aβ)]-

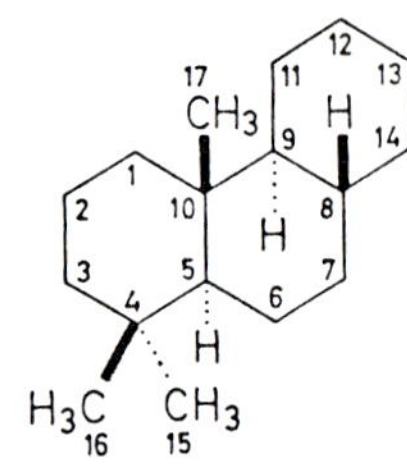

$C_{17}H_{30}$
Beilstein: 3-6.2098* · Syst.-Nr. 461
CAS-Reg.No.: 471-78-3
IUPAC: F-(27)

–, 13-ethyl-13-methyl- = Pimarane
–, 13-isopropyl- = Abietane
–, 14-isopropyl- = *Totarane*

Porphine **Porphin**

= Porphyrin

Porphyrin **Porphyrin**

syn: Porphine
sys[CA]: 21*H*,23*H*-Porphine

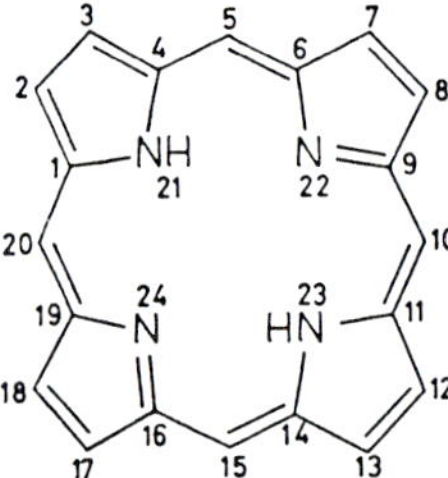

$C_{20}H_{14}N_4$
Beilstein: 4-26.1900 · Syst.-Nr. 4031
CAS-Reg.No.: 101-60-0
IUPAC: TP-1

–, 2,3-dihydro- = *Chlorin*
–, 5,22-dihydro- = *Phlorin*
–, 5,10,15,20,22,24-hexahydro- = Porphyrinogen
–, 7,8,17,18-tetrahydro- = Bacteriochlorin

Porphyrinogen **Porphyrinogen**

sys[CA]: 21*H*,23*H*-Porphine, 5,10,15,20,22,24-hexahydro-

$C_{20}H_{20}N_4$
Beilstein · Syst.-Nr. 4028
CAS-Reg.No.: 4396-11-6
IUPAC: TP-4.1

Pregnane **Pregnan**

$C_{21}H_{36}$
Beilstein: 4-5.1215 · Syst.-Nr. 472
CAS-Reg.No.: 24909-91-9
IUPAC: 2S-2.3(20)

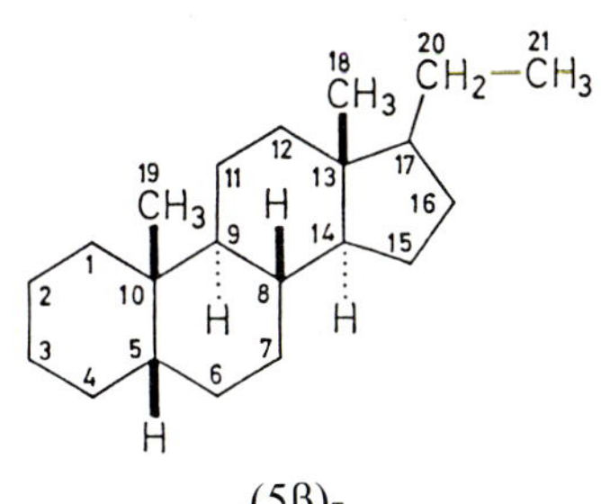

(5β)-

(5α)-Pregnane [syn: Allopregnane]
CAS-Reg.No.: 641-85-0
(5β)-Pregnane
CAS-Reg.No.: 481-26-5
(5α,14β,17β*H*)-Pregnane [syn: Diginane]
CAS-Reg.No.: 78964-23-5

Primverose **Primverose**

= D-Glucose, O^6-β-D-xylopyranosyl-

Progesterone **Progesteron**

= Pregn-4-ene-3,20-dione

Proline **Prolin**

$C_5H_9NO_2$
Beilstein: 4-22.8 · Syst.-Nr. 3244
CAS-Reg.No.: 7005-20-1
IUPAC: 3AA-2.2.2

D-Proline
Beilstein: 4-22.8
CAS-Reg.No.: 344-25-2
L-Proline
Beilstein: 4-22.8
CAS-Reg.No.: 147-85-3
DL-Proline
Beilstein: 4-22.12
CAS-Reg.No.: 609-36-9

Propane **Propan**

C_3H_8
Beilstein: 4-1.176 · Syst.-Nr. 10
CAS-Reg.No.: 74-98-6
IUPAC: A-1.1

$CH_3-CH_2-CH_3$

Propiolic acid **Propiolsäure**

sys[CA]: 2-Propynoic acid

$C_3H_2O_2$

Beilstein: 4-2.1687 · Syst.-Nr. 164

CAS-Reg.No.: 471-25-0

IUPAC: C-404.1

$HC \equiv CH - CO - OH$

Propionic acid **Propionsäure**

sys[CA]: Propanoic acid

$C_3H_6O_2$

Beilstein: 4-2.695 · Syst.-Nr. 162

CAS-Reg.No.: 79-09-4

IUPAC: C-404.1

$CH_3 - CH_2 - CO - OH$

–, 2,2-dimethyl- = *Pivalic acid*

Prostane **Prostan**

$C_{20}H_{40}$

Beilstein · Syst.-Nr. 452

CAS-Reg.No.: 36413-57-7

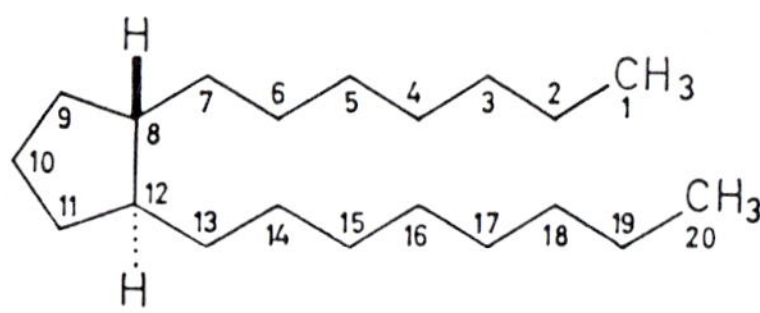

Protocatechuic acid **Protocatechusäure**

sys[CA]: Benzoic acid, 3,4-dihydroxy-

$C_7H_6O_4$

Beilstein: 4-10.1459 · Syst.-Nr. 1105

CAS-Reg.No.: 99-50-3

IUPAC: C-411.1

Protoporphyrin **Protoporphyrin**

syn: Protoporphyrin IX

sys[CA]: 21*H*,23*H*-Porphine-2,18-dipropanoic acid, 7,12-diethenyl-3,8,13,17-tetramethyl-

$C_{34}H_{34}N_4O_4$

Beilstein: 4-26.3042 · Syst.-Nr. 4173

CAS-Reg.No.: 553-12-8

IUPAC: TP-2.1

Protostane **Protostan**

= Dammarane, (8α,9β,13α,14β)-

Psicofuranose **Psicofuranose**

$C_6H_{12}O_6$
Beilstein · Syst.-Nr. 145
IUPAC: Carb-10/18

α-D-Psicofuranose
CAS-Reg.No.: 41847-06-7
β-D-Psicofuranose
CAS-Reg.No.: 470-24-6

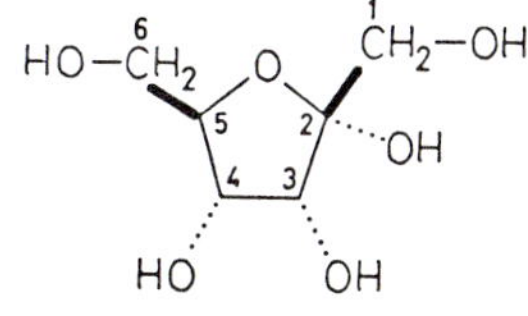

α-D-

Psicopyranose **Psicopyranose**

$C_6H_{12}O_6$
Beilstein · Syst.-Nr. 145
IUPAC: Carb-10/18

α-D-Psicopyranose
CAS-Reg.No.: 38029-84-4
β-D-Psicopyranose
CAS-Reg.No.: 40461-85-6
α-L-Psicopyranose
CAS-Reg.No.: 41847-53-4

α-D-

Psicose **Psicose**

$C_6H_{12}O_6$
Beilstein · Syst.-Nr. 145
CAS-Reg.No.: 23140-52-5
IUPAC: Carb-10

D-Psicose
Beilstein: 4-1.4400
CAS-Reg.No.: 551-68-8
L-Psicose
Beilstein: 4-1.4401
CAS-Reg.No.: 16354-64-6

D-

Pteridine **Pteridin**

$C_6H_4N_4$
Beilstein: 4-26.1770 · Syst.-Nr. 4022
CAS-Reg.No.: 91-18-9
IUPAC: B-2.11(33)

Pteroic acid **Pteroinsäure**

sys[CA]: Benzoic acid, 4-[[(2-amino-1,4-dihydro-4-oxo-6-pteridinyl)-methyl]amino]-

$C_{14}H_{12}N_6O_3$
Beilstein: 4-26.3942 · Syst.-Nr. 4179
CAS-Reg.No.: 119-24-4
IUPAC: Fol-2

Purine **Purin**

sys[CA]: 1*H*-Purine

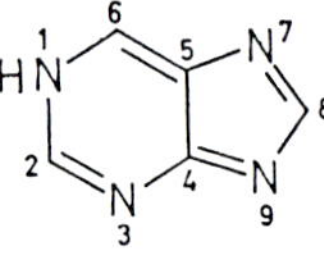

$C_5H_4N_4$
Beilstein: 4-26.1736 · Syst.-Nr. 4019
CAS-Reg.No.: 120-73-0
IUPAC: B-2.11(24)

9*H*-Purine
CAS-Reg.No.: 51953-03-8

Pyran **Pyran**

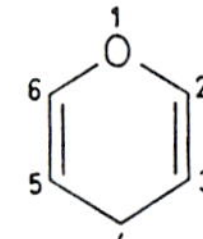

C_5H_6O
Beilstein: 5-17/1.321 · Syst.-Nr. 2364
IUPAC: B-2.11(6)

2*H*-Pyran
CAS-Reg.No.: 289-66-7
4*H*-Pyran
CAS-Reg.No.: 289-65-6

Pyranthrene **Pyranthren**

$C_{30}H_{16}$
Beilstein: 4-5.2915 · Syst.-Nr. 497
CAS-Reg.No.: 191-13-9
IUPAC: A-21.2(34)

Pyrazine **Pyrazin**

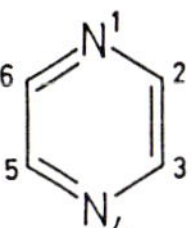

$C_4H_4N_2$
Beilstein: 4-23.899 · Syst.-Nr. 3469
CAS-Reg.No.: 290-37-9
IUPAC: B-2.11(16)

Pyrazole **Pyrazol**

sys[CA]: 1*H*-Pyrazole

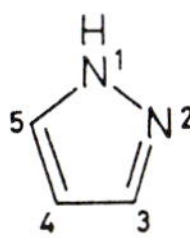

$C_3H_4N_2$
Beilstein: 4-23.550 · Syst.-Nr. 3463
IUPAC: B-2.11(14)

1*H*-Pyrazole
CAS-Reg.No.: 288-13-1
3*H*-Pyrazole
CAS-Reg.No.: 288-12-0
4*H*-Pyrazole
CAS-Reg.No.: 288-11-9

–, dihydro- = Pyrazoline

Pyrazolidine **Pyrazolidin**

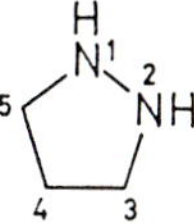

$C_3H_8N_2$
Beilstein: 4-23.4 · Syst.-Nr. 3460 A
CAS-Reg.No.: 504-70-1
IUPAC: B-2.12(7)

Pyrazoline **Pyrazolin**

sys[B]: Pyrazol, Dihydro-
[CA]: 3*H*-Pyrazole, 4,5-dihydro-

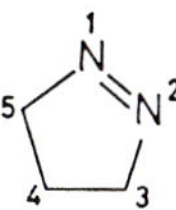

$C_3H_6N_2$
Beilstein: 4-23.454 · Syst.-Nr. 3461
CAS-Reg.No.: 2721-43-9
IUPAC: B-2.12(8)

Pyrene **Pyren**

$C_{16}H_{10}$, formula A [former numbering: formula B]
Beilstein: 4-5.2467 · Syst.-Nr. 487
CAS-Reg.No.: 129-00-0
IUPAC: A-21.1(18)

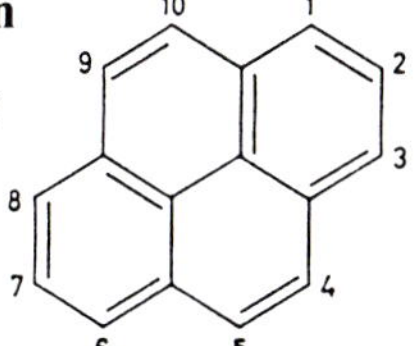

A

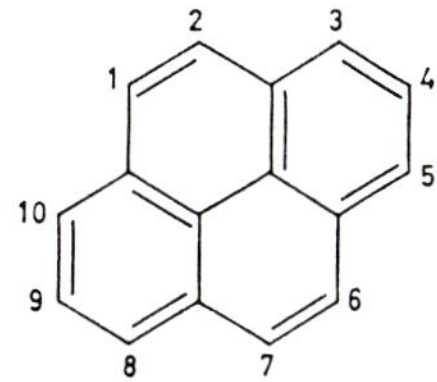

B

Pyridazine **Pyridazin**

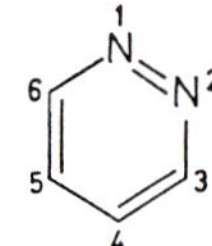

$C_4H_4N_2$
Beilstein: 4-23.889 · Syst.-Nr. 3469
CAS-Reg.No.: 289-80-5
IUPAC: B-2.11(18)

Pyridine **Pyridin**

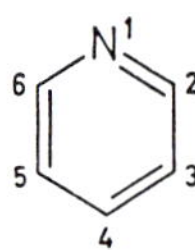

C_5H_5N
Beilstein: 4-20.2205 · Syst.-Nr. 3051
CAS-Reg.No.: 110-86-1
IUPAC: B-2.11(15)

Pyridoxal **Pyridoxal**

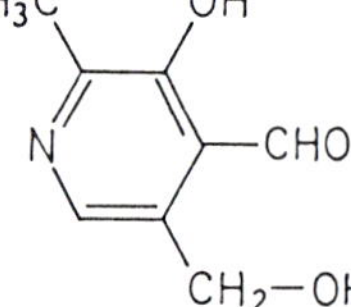

$C_8H_9NO_3$
Beilstein: 4-21.6417 · Syst.-Nr. 3240
CAS-Reg.No.: 66-72-8
IUPAC: VB_6-7.2

Pyridoxamine **Pyridoxamin**

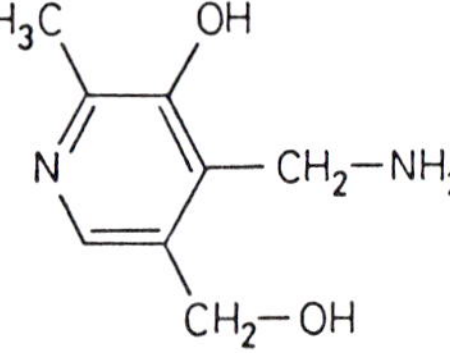

$C_8H_{12}N_2O_2$
Beilstein: 4-22.6063 · Syst.-Nr. 3426
CAS-Reg.No.: 85-87-0
IUPAC: VB_6-7.3

Pyridoxine **Pyridoxin**

syn: Vitamin B_6, PN, Pyridoxol, Adermine
sys[CA]: 3,4-Pyridinedimethanol, 5-hydroxy-6-methyl-

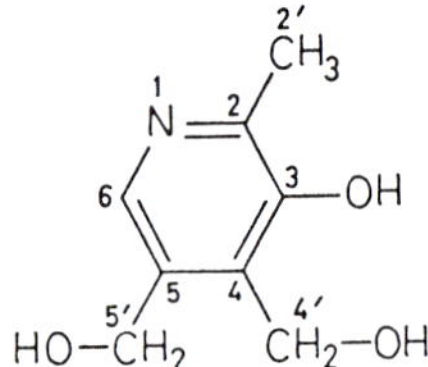

$C_8H_{11}NO_3$
Beilstein: 4-21.2509 · Syst.-Nr. 3157
CAS-Reg.No.: 65-23-6
IUPAC: VB_6-7.1

Pyrimidine **Pyrimidin**

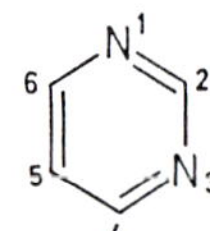

$C_4H_4N_2$
Beilstein: 4-23.892 · Syst.-Nr. 3469
CAS-Reg.No.: 289-95-2
IUPAC: B-2.11(17)

Pyrindine **Pyrindin**

C_8H_9N
Beilstein · Syst.-Nr. 3069

7*H*-[1]

1*H*-[1]Pyrindine
CAS-Reg.No.: 270-88-2
2*H*-[1]Pyrindine
CAS-Reg.No.: 270-90-6
4*H*-[1]Pyrindine
CAS-Reg.No.: 270-89-3
4a*H*-[1]Pyrindine
CAS-Reg.No.: 270-87-1
5*H*-[1]Pyrindine
CAS-Reg.No.: 270-91-7
7*H*-[1]Pyrindine
CAS-Reg.No.: 270-92-8

1*H*-[2]Pyrindine
CAS-Reg.No.: 270-56-4
2*H*-[2]Pyrindine
CAS-Reg.No.: 270-58-6
3*H*-[2]Pyrindine
CAS-Reg.No.: 270-57-5
4a*H*-[2]Pyrindine
CAS-Reg.No.: 270-55-3
5*H*-[2]Pyrindine
CAS-Reg.No.: 270-60-0
6*H*-[2]Pyrindine
CAS-Reg.No.: 270-59-7

Pyrocatechol **Brenzcatechin**
sys[CA]: 1,2-Benzenediol

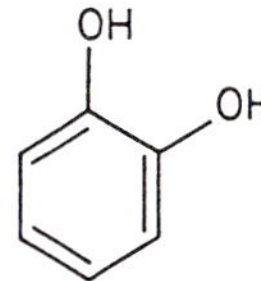

$C_6H_6O_2$
Beilstein: 4-6.5557 · Syst.-Nr. 553
CAS-Reg.No.: 120-80-9
IUPAC: C-202.2

Pyrogallol **Pyrogallol**
sys[CA]: 1,2,3-Benzenetriol

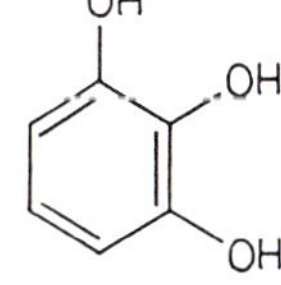

$C_6H_6O_3$
Beilstein: 4-6.7327 · Syst.-Nr. 578
CAS-Reg.No.: 87-66-1
IUPAC: C-202.2

Pyrrochlorin **Pyrrochlorin**

syn: Mesopyrrochlorin

sys[CA]: 21*H*,23*H*-Porphine-2-propanoic acid, 8,13-diethyl-2,3-dihydro-3,7,12,17-tetramethyl-, (2*S*-*trans*)-

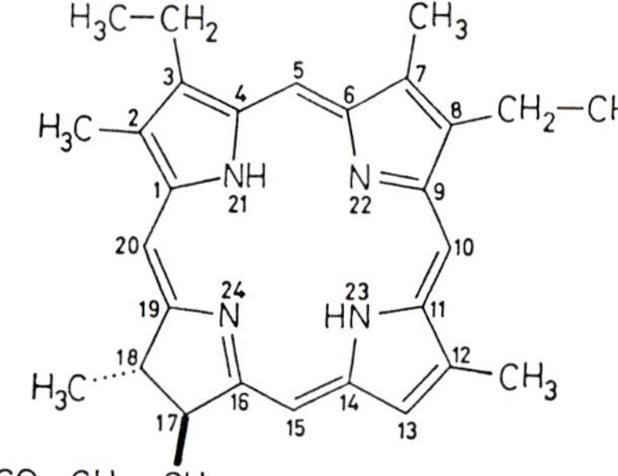

$C_{31}H_{36}N_4O_2$
Beilstein: 4-26.2942 · Syst.-Nr. 4173
CAS-Reg.No.: 26791-77-5
IUPAC: TP-4.3.1

Pyrrole **Pyrrol**

sys[CA]: 1*H*-Pyrrole

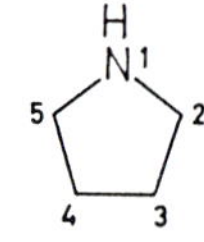

C_4H_5N
Beilstein: 4-20.2072 · Syst.-Nr. 3048
IUPAC: B-2.11(12)

1*H*-Pyrrole
CAS-Reg.No.: 109-97-7
2*H*-Pyrrole
CAS-Reg.No.: 287-95-6
3*H*-Pyrrole
CAS-Reg.No.: 287-97-8

–, dihydro- = Pyrroline

Pyrrolidine **Pyrrolidin**

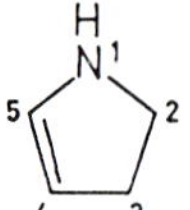

C_4H_9N
Beilstein: 4-20.61 · Syst.-Nr. 3037 A
CAS-Reg.No.: 123-75-1
IUPAC: B-2.12(3)

Pyrroline **Pyrrolin**

sys[B+CA]: Pyrrole, dihydro-

C_4H_7N
Beilstein · Syst.-Nr. 3047
CAS-Reg.No.: 5724-81-2
IUPAC: B-2.12(4)

Pyrroporphyrin **Pyrroporphyrin**

syn: Pyrroporphyrin XV

sys[CA]: 21*H*,23*H*-Porphine-2-propanoic acid, 8,13-diethyl-3,7,12,17-tetramethyl-

$C_{31}H_{34}N_4O_2$

Beilstein: 4-26.2951 · Syst.-Nr. 4173

CAS-Reg.No.: 644-00-8

IUPAC: TP-2.1

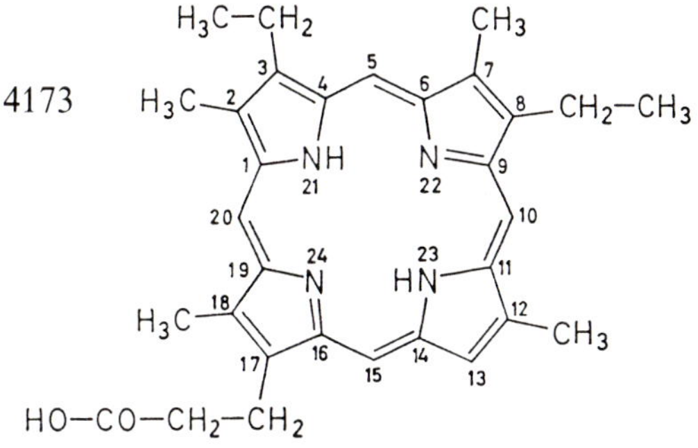

Pyruvic acid **Brenztraubensäure**

sys[B]: Propionsäure, 2-Oxo-

[CA]: Propanoic acid, 2-oxo-

$C_3H_4O_3$

Beilstein: 4-3.1505 · Syst.-Nr. 279

CAS-Reg.No.: 127-17-3

IUPAC: C-416.3

$CH_3-CO-CO-OH$

Quercitol **Quercit**

sys[CA]: D-*chiro*-Inositol, 2-deoxy-

$C_6H_{12}O_5$

Beilstein: 4-6.7883 · Syst.-Nr. 603

CAS-Reg.No.: 488-73-3

IUPAC: I-2

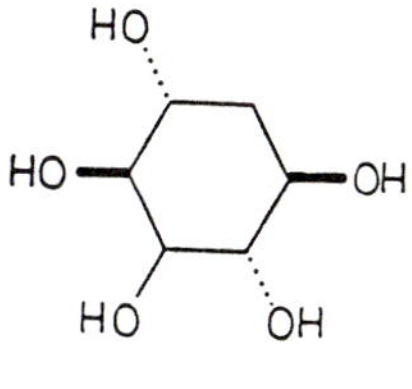

Quinazoline **Chinazolin**

$C_8H_6N_2$

Beilstein: 4-23.1221 · Syst.-Nr. 3480

CAS-Reg.No.: 253-82-7

IUPAC: B-2.11(31)

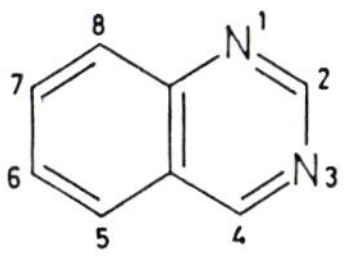

Quindoline **Chindolin**

sys[B]: Indolo[3,2-*b*]chinolin

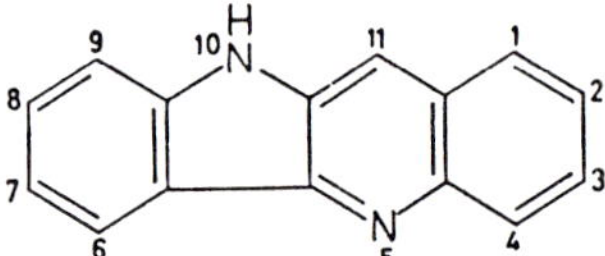

$C_{15}H_{10}N_2$
Beilstein: 4-23.1869 · Syst.-Nr. 3489

1*H*-Quindoline
CAS-Reg.No.: 42375-21-3
11*H*-Quindoline
CAS-Reg.No.: 243-57-2

Quinic acid **Chinasäure**

sys[CA]: Cyclohexanecarboxylic acid, 1,3,4,5-tetrahydroxy-, (1α,3α,4α,5β)-

$C_7H_{12}O_6$
Beilstein · Syst.-Nr. 1159
CAS-Reg.No.: 36413-60-2
IUPAC: I-8(23)

(−)-Quinic acid
Beilstein: 4-10.2257
CAS-Reg.No.: 77-95-2
(±)-Quinic acid
Beilstein: 4-10.2258

Quinindoline **Chinindolin**

sys[B]: Indolo[2,3-*b*]chinolin

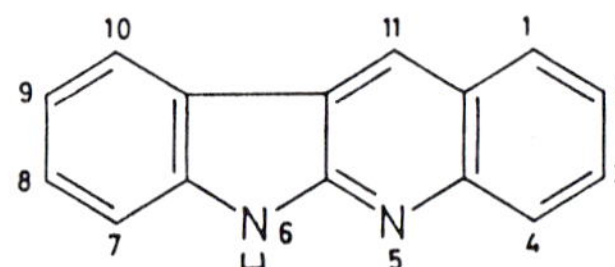

$C_{15}H_{10}N_2$
Beilstein: 4-23.1868 · Syst.-Nr. 3489

1*H*-Quinindoline
CAS-Reg.No.: 42375-22-4
5*H*-Quinindoline
CAS-Reg.No.: 243-38-9

Quinoline **Chinolin**

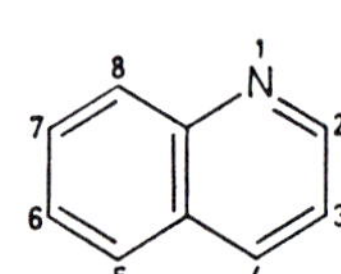

C_9H_7N
Beilstein: 4-20.3334 · Syst.-Nr. 3077
CAS-Reg.No.: 91-22-5
IUPAC: B-2.11(27)

Quinolizidine **Chinolizidin**

= Quinolizine, octahydro-

Quinolizine **Chinolizin**

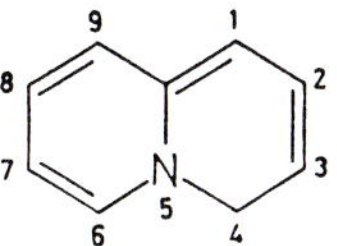

C_9H_9N
Beilstein · Syst.-Nr. 3070
CAS-Reg.No.: 255-58-3
IUPAC: B-2.11(25)

–, octahydro- = *Quinolizidine*

Quinovose **Chinovose**

= D-Glucose, 6-deoxy-

Quinoxaline **Chinoxalin**

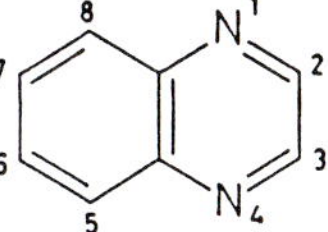

$C_8H_6N_2$
Beilstein: 4-23.1226 · Syst.-Nr. 3480
CAS-Reg.No.: 91-19-0
IUPAC: B-2.11(30)

Quinuclidine **Chinuclidin**
sys[CA]: 1-Azabicyclo[2.2.2]octane

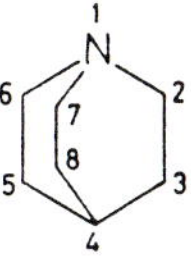

$C_7H_{13}N$
Beilstein: 4-20.1966 · Syst.-Nr. 3047
CAS-Reg.No.: 100-76-5
IUPAC: B-2.12(13)

Resorcinol **Resorcin**
sys[CA]: 1,3-Benzenediol

OH

OH

$C_6H_6O_2$
Beilstein: 4-6.5658 · Syst.-Nr. 554
CAS-Reg.No.: 108-46-3
IUPAC: C-202.2

Retinal **Retinal**

syn: Vitamin A aldehyde, Retinene

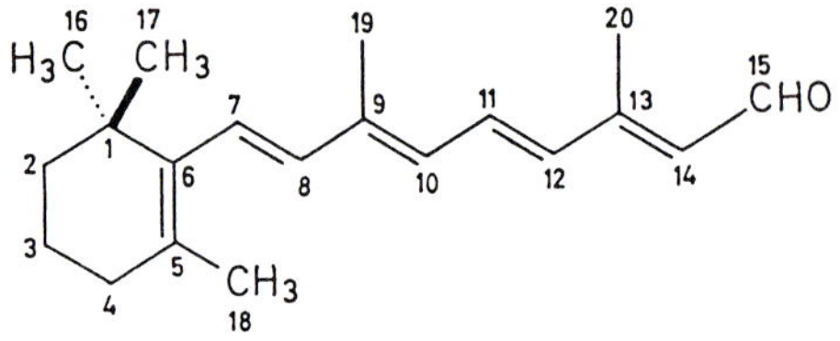

$C_{20}H_{28}O$
Beilstein: 4-7.1253 · Syst.-Nr. 648
CAS-Reg.No.: 116-31-4
IUPAC: Ret-2.2

Retinene **Retinen**

= Retinal

Retinoic acid **Retinsäure**

syn: Vitamin A acid, Tretinoin

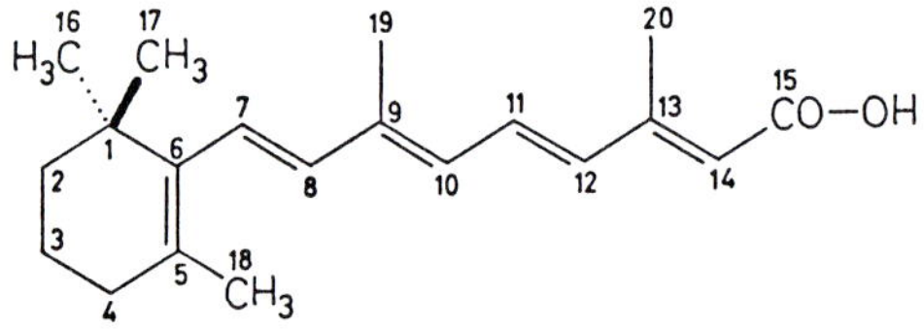

$C_{20}H_{28}O_2$
Beilstein: 4-9.2387 · Syst.-Nr. 950
CAS-Reg.No.: 302-79-4
IUPAC: Ret-2.3

Retinol **Retinol**

syn: Vitamin A alcohol, Vitamin A, Axerophthol, Axerol

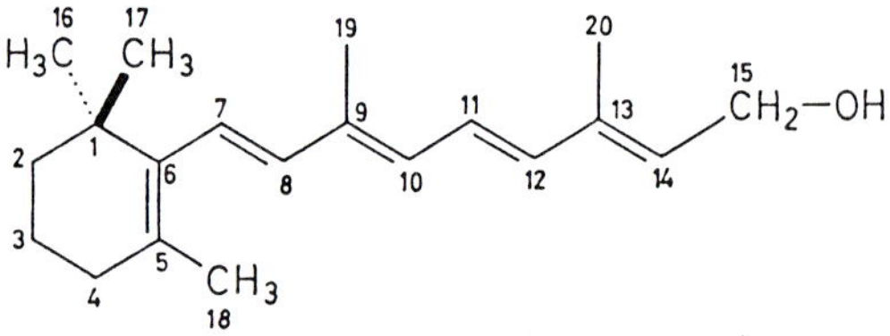

$C_{20}H_{30}O$
Beilstein: 4-6.4133 · Syst.-Nr. 535
CAS-Reg.No.: 68-26-8
IUPAC: Ret-2.1

Retinsäure = Retinoic acid

Rhamnose **Rhamnose**

= Mannose, 6-deoxy-

Rheadan **Rheadan**

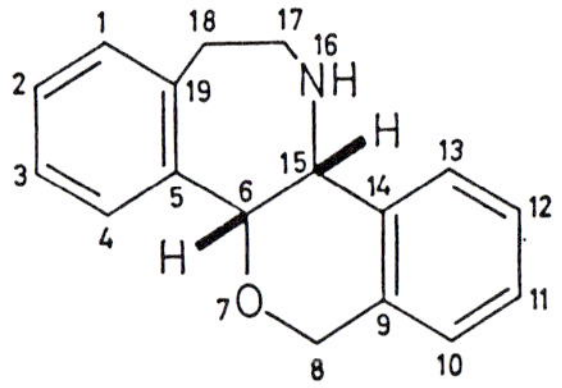

$C_{17}H_{17}NO$
Beilstein: 4-27.6805* · Syst.-Nr. 4199
CAS-Reg.No.: 34437-70-2
IUPAC: F-2

Rhodanine **Rhodanin**

sys[B]: Thiazolidin-4-on, 2-Thioxo-
[CA]: 4-Thiazolidinone, 2-thioxo-

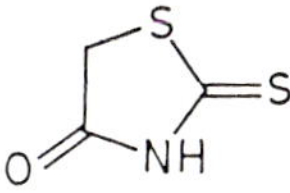

$C_3H_3NOS_2$
Beilstein: 4-27.3188 · Syst.-Nr. 4298
CAS-Reg.No.: 141-84-4
IUPAC: C-548

Rhodinose **Rhodinose**

= Hexanal, *threo*-4,5-dihydroxy-

Rhodochlorin **Rhodochlorin**

sys[CA]: 21*H*,23*H*-Porphine-2-propanoic acid, 18-carboxy-8,13-diethyl-17,18-dihydro-3,7,12,-17-tetramethyl-

$C_{32}H_{36}N_4O_4$
Beilstein: 4-26.2985 · Syst.-Nr. 4173
IUPAC: TP-4.3.1

Rhodoporphyrin **Rhodoporphyrin**

syn: Rhodoporphyrin XV
sys[CA]: 21*H*,23*H*-Porphine-2-propanoic acid, 18-carboxy-8,13-diethyl-3,7,12,17-tetramethyl-

$C_{32}H_{34}N_4O_4$
Beilstein: 4-26.3005 · Syst.-Nr. 4173
CAS-Reg.No.: 531-12-4
IUPAC: TP-2.1

Rhodymenabiose **Rhodymenabiose**

= D-Xylose, O^3-β-D-xylopyranosyl-

Riboflavin **Riboflavin**

syn: Lactoflavin, Vitamin B_2

$C_{17}H_{20}N_4O_6$
Beilstein: 4-26.2542 · Syst.-Nr. 4142
CAS-Reg.No.: 83-88-5
IUPAC: M-5

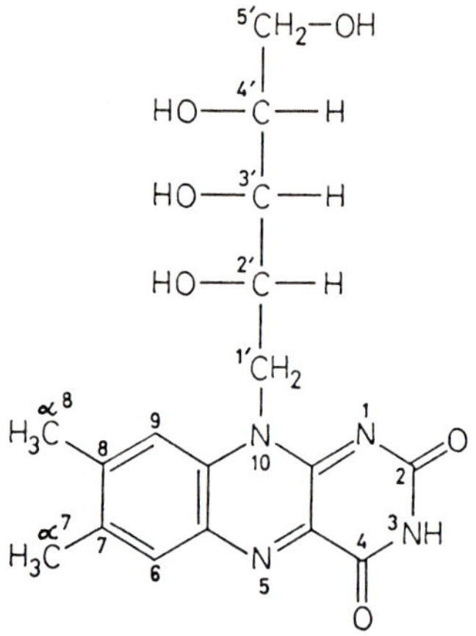

Ribofuranose **Ribofuranose**

$C_5H_{10}O_5$
Beilstein · Syst.-Nr. 133
CAS-Reg.No.: 15761-67-8
IUPAC: Carb-5/18

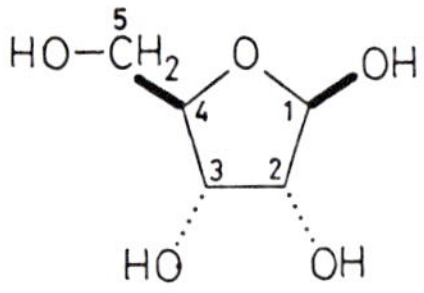

α-D-Ribofuranose
Beilstein: 4-1.4211
CAS-Reg.No.: 32445-75-3
β-D-Ribofuranose
Beilstein: 4-1.4211
CAS-Reg.No.: 36468-53-8
α-L-Ribofuranose
CAS-Reg.No.: 41546-20-7
β-L-Ribofuranose
CAS-Reg.No.: 41546-19-4

β-D-

Ribopyranose **Ribopyranose**

$C_5H_{10}O_5$
Beilstein · Syst.-Nr. 133
IUPAC: Carb-5/18

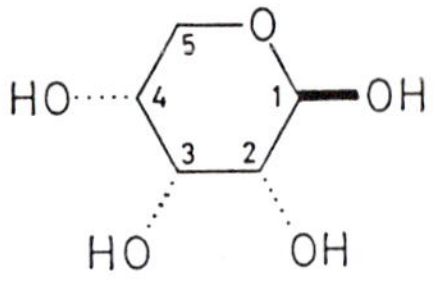

α-D-Ribopyranose
Beilstein: 4-1.4211
CAS-Reg.No.: 7296-59-5
β-D-Ribopyranose
Beilstein: 4-1.4211
CAS-Reg.No.: 7296-60-8
α-L-Ribopyranose
CAS-Reg.No.: 7296-61-9
β-L-Ribopyranose
CAS-Reg.No.: 7296-62-0

β-D-

Ribose **Ribose**

$C_5H_{10}O_5$
Beilstein · Syst.-Nr. 133
CAS-Reg.No.: 34466-20-1
IUPAC: Carb-5

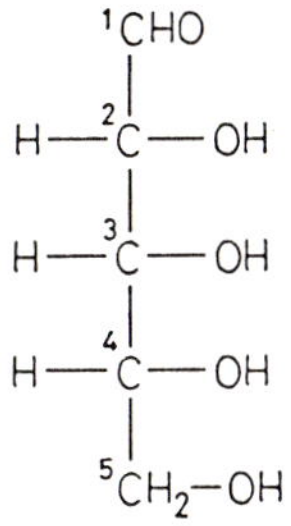

–, 2-hydroxymethyl- = *Hamamelose*

D-Ribose
Beilstein: 4-1.4212
CAS-Reg.No.: 50-69-1
L-Ribose
Beilstein: 4-1.4214
CAS-Reg.No.: 24259-59-4
DL-Ribose
Beilstein: 4-1.4215
CAS-Reg.No.: 55058-43-0

D-

Rimuane **Rimuan**

= Rosane, (13*S*)-(10α)-

Rodiasan **Rodiasan**

$C_{32}H_{30}N_2O$
Beilstein · Syst.-Nr. 4502
IUPAC: F-2

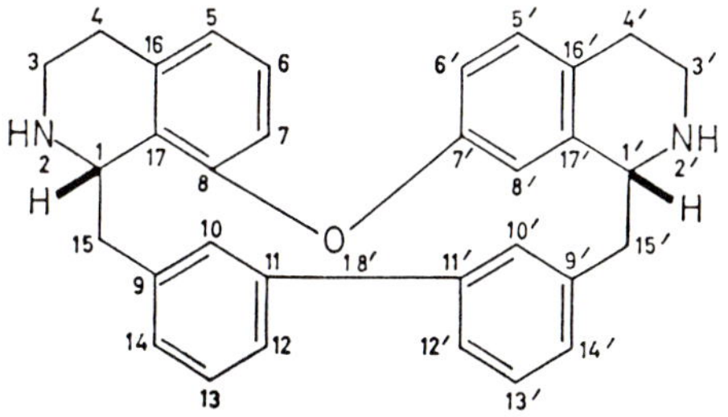

Rodiasine **Rodiasin**

= Rodiasan, 12′-hydroxy-6,7,12,6′-tetramethoxy-2,2′-dimethyl-

---

Rosane **Rosan**

sys[CA]: Phenanthrene, 7-ethyl-tetradecahydro-1,1,4b,7-tetramethyl-, [4a*R*-(4aα,4bβ,7β,-8aα,10aα)]-

$C_{20}H_{36}$
Beilstein: 4-17.4777* · Syst.-Nr. 461
CAS-Reg.No.: 6812-82-4
IUPAC: F-2

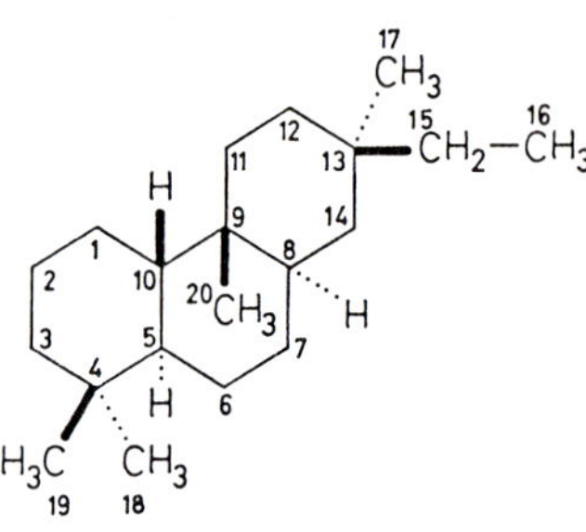

–, (13*S*)-(10α)- = *Rimuane*

Rubicene **Rubicen**

$C_{26}H_{14}$
Beilstein: 4-5.2864 · Syst.-Nr. 495
CAS-Reg.No.: 197-61-5
IUPAC: A-21.2(29)

Rugulosin **Rugulosin**

$C_{30}H_{22}O_{10}$
Beilstein: 4-8.3766 · Syst.-Nr. 890
CAS-Reg.No.: 23537-16-8

Rutinose **Rutinose**

= D-Glucose, O^6-[α-L-6-deoxy-mannopyranosyl]-

Saccharose = Sucrose

Safrole **Safrol**

sys[B]: Benzo[1,3]dioxol, 5-Allyl-
[CA]: 1,3-Benzodioxole, 5-(2-propenyl)-

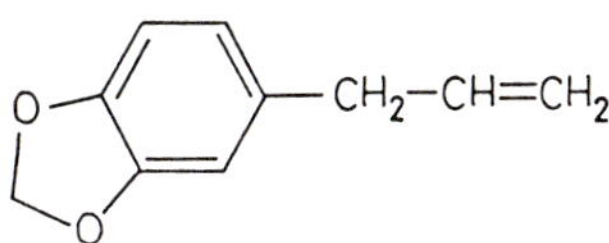

$C_{10}H_{10}O_2$
Beilstein: 4-19.275 · Syst.-Nr. 2673
CAS-Reg.No.: 94-59-7
IUPAC: C-331.3

Salicylic acid **Salicylsäure**

sys[B+CA]: Benzoic acid, 2-hydroxy-

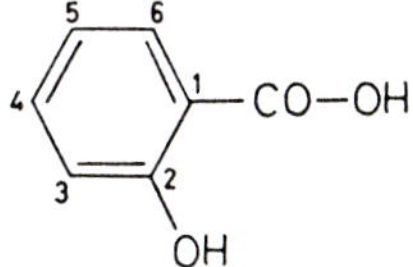

$C_7H_6O_3$
Beilstein: 4-10.125 · Syst.-Nr. 1057
CAS-Reg.No.: 69-72-7
IUPAC: C-411.1

Sapogenin **Sapogenin**

= Spirostan, (25*S*)-

Sapphyrin **Sapphyrin**

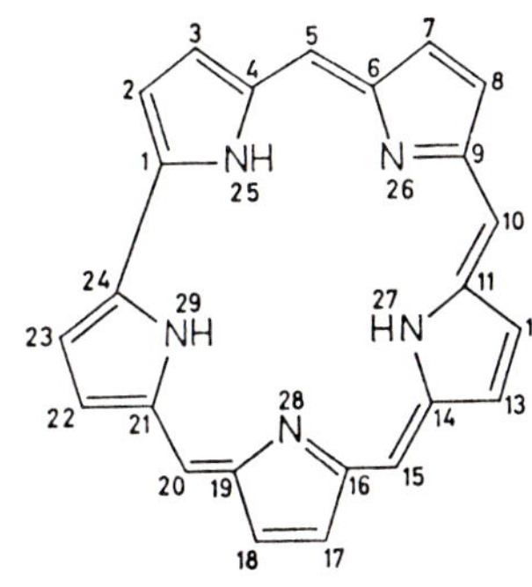

$C_{24}H_{17}N_5$
Beilstein · Syst.-Nr. 4187
IUPAC: TP-5.4

Sarcosine **Sarkosin**

sys[B+CA]:Glycine, *N*-methyl-

$C_3H_7NO_2$
Beilstein: 4-4.2363 · Syst.-Nr. 364
CAS-Reg.No.: 107-97-1
IUPAC: C-421.1; 3AA-Appendix

$CH_3-NH-CH_2-CO-OH$

Sarmentose **Sarmentose**

= *xylo*-2,6-Dideoxy-hexose, O^3-methyl-

Sarpagane **Sarpagan**

$C_{19}H_{22}N_2$
Beilstein: 4-23.2833* · Syst.-Nr. 3487
CAS-Reg.No.: 3921-76-4
IUPAC: F-2

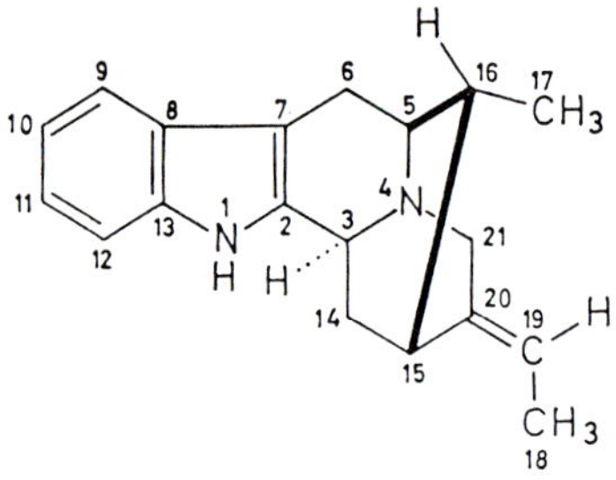

Sarsasapogenin **Sarsasapogenin**

= Spirostan-3β-ol, (25*S*)-(5β)-

Scillabiose **Scillabiose**

= L-Mannose, O^4-β-D-glucopyranosyl-6-deoxy-

Scillarenin **Scillarenin**

= Bufa-4,20,22-trienolide, 3β,14-dihydroxy-(14β)-

Sebacic acid **Sebacinsäure**

= Decandioic acid

Securinan **Securinan**

$C_{13}H_{17}NO$
Beilstein · Syst.-Nr. 4195
CAS-Reg.No.: 34479-49-7

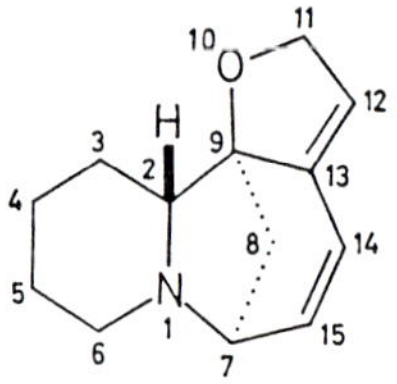

Sedoheptulose **Sedoheptulose**

sys[B+CA]: D-*altro*-[2]Heptulose

$C_7H_{14}O_7$
Beilstein: 4-1.4446 · Syst.-Nr. 149
CAS-Reg.No.: 3019-74-7
IUPAC: Carb-10

Semicarbazide **Semicarbazid**

sys[CA]: Hydrazinecarboxamide

CH_5N_3O
Beilstein: 4-3.177 · Syst.-Nr. 209 B
CAS-Reg.No.: 57-56-7
IUPAC: C-981.1

$H_2N-CO-NH-NH_2$

Senecioic acid **Seneciosäure**

= Crotonic acid, 3-methyl-

Senecionan **Senecionan**

$C_{18}H_{29}NO_2$
Beilstein: 4-21.250* · Syst.-Nr. 4401
CAS-Reg.No.: 34486-91-4
IUPAC: F-2

Serine **Serin**

$C_3H_7NO_3$

$HO-CH_2-CH(NH_2)-CO-OH$

Beilstein: 4-4.3118 · Syst.-Nr. 376
CAS-Reg.No.: 6898-95-9
IUPAC: 3AA-1

D-Serine
Beilstein: 4-4.3118
CAS-Reg.No.: 312-84-5
L-Serine
Beilstein: 4-4.3118
CAS-Reg.No.: 56-45-1
DL-Serine
Beilstein: 4-4.3119
CAS-Reg.No.: 302-84-1

Serratane **Serratan**

= Gammacerane, *C*(14a)-homo-27-nor-

Shionane **Shionan**

= Lupane, *D:A*-friedo-18,19-seco-

Simiarane **Simiaran**

= Gammacerane, *D:B*-friedo-*B′:A′*-neo-

Solabiose **Solabiose**

= D-Galactose, O^3-β-D-glucopyranosyl-

Solanidane **Solanidan**

$C_{27}H_{45}N$
Beilstein: 4-20.3332 · Syst.-Nr. 3073
CAS-Reg.No.: 562-08-3
IUPAC: F-2

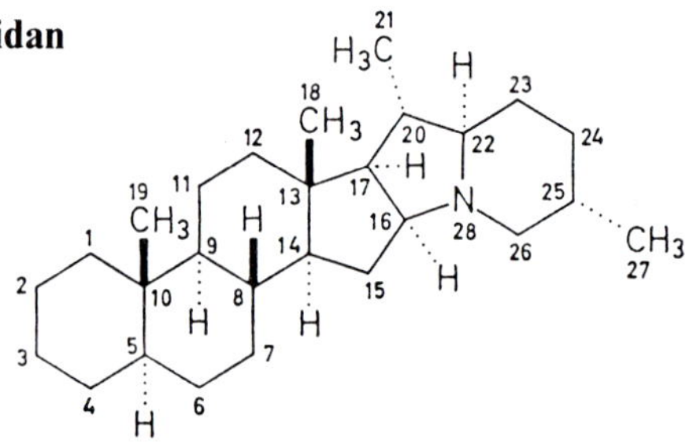

Solanidanine **Solanidanin**

= Solanidane

Solasodane **Solasodan**

= Spirosolane, (22*R*,25*R*)-(5α)-

Sophoridane **Sophoridan**

= Matridine, (5β)-

Sophorose **Sophorose**

= D-Glucose, O^2-β-D-glucopyranosyl-

Sorbofuranose **Sorbofuranose**

$C_6H_{12}O_6$
Beilstein · Syst.-Nr. 145
IUPAC: Carb-10/18

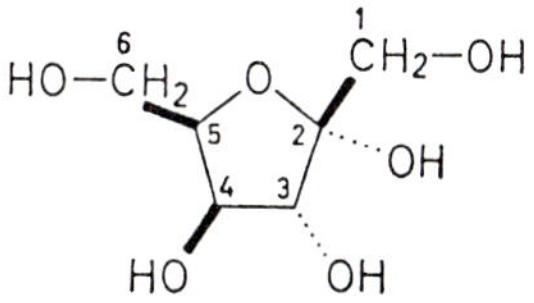

α-D-Sorbofuranose
CAS-Reg.No.: 41847-03-4
β-D-Sorbofuranose
CAS-Reg.No.: 42744-42-3
α-L-Sorbofuranose
CAS-Reg.No.: 36468-68-5
β-L-Sorbofuranose
CAS-Reg.No.: 41847-02-3

α-D-

Sorbopyranose **Sorbopyranose**

$C_6H_{12}O_6$
Beilstein · Syst.-Nr. 145
IUPAC: Carb-10/18

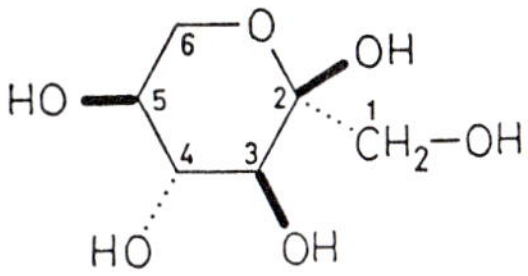

α-D-Sorbopyranose
CAS-Reg.No.: 41847-56-7
α-L-Sorbopyranose
Beilstein: 4-1.4412
CAS-Reg.No.: 470-15-5
β-L-Sorbopyranose
CAS-Reg.No.: 35831-80-2

α-L-

Sorbose **Sorbose**

$C_6H_{12}O_6$
Beilstein · Syst.-Nr. 145
CAS-Reg.No.: 3615-39-2
IUPAC: Carb-10

D-Sorbose
Beilstein: 4-1.4411
CAS-Reg.No.: 3615-56-3
L-Sorbose
Beilstein: 4-1.4412
CAS-Reg.No.: 87-79-6
DL-Sorbose
CAS-Reg.No.: 65732-90-3

$^1CH_2$—OH
^{2}CO
H—^{3}C—OH
HO—^{4}C—H
H—^{5}C—OH
$^6CH_2$—OH

D-

Sparteine **Spartein**
sys[CA]: 7,14-Methano-2H,6H-dipyrido-[1,2-*a*:′,2′-*e*][1,5]diazocine, dodecahydro-, [7*S*-(7α,7aα,-14α,14aβ)]-

$C_{15}H_{26}N_2$
Beilstein: 4-23.957 · Syst.-Nr. 3469
CAS-Reg.No.: 90-39-1

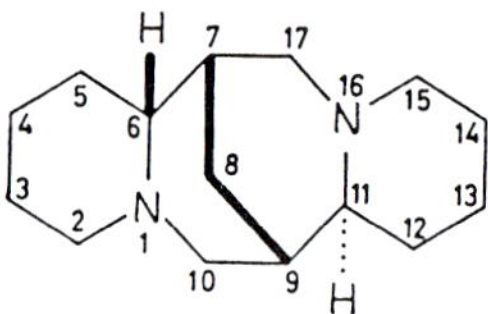

Sphinganine **Sphinganin**
sys[CA]: 1,3-Octadecanediol, 2-amino-, [*R*-(*R**,*S**)]-

$C_{18}H_{39}NO_2$
Beilstein: 4-4.1887 · Syst.-Nr. 356
CAS-Reg.No.: 764-22-7
IUPAC: Lip-1.8

$^1CH_2$—OH
H—^{2}C—NH_2
H—^{3}C—OH
$[CH_2]_{14}$
$^{18}CH_3$

Sphingosine **Sphingosin**
sys[CA]: 4-Octadecene-1,3-diol, 2-amino-, [*R*-[*R**,*S**]-(*E*)]]-

$C_{18}H_{37}NO_2$
Beilstein: 4-4.1894 · Syst.-Nr. 356
CAS-Reg.No.: 123-78-4
IUPAC: Lip-1.11

CH_2—OH
H—C—NH_2
H—C—OH
H—C=C—H
$[CH_2]_{12}$
CH_3

Spirosolane **Spirosolan**

$C_{27}H_{45}NO$
Beilstein: 4-27.1138 · Syst.-Nr. 4195
CAS-Reg.No.: 34471-97-1
IUPAC: F-2

–, (5α)- = *Tomatidane, Tomatanine*
–, (22*R*,25*R*)-(5α)- = *Solasodane*

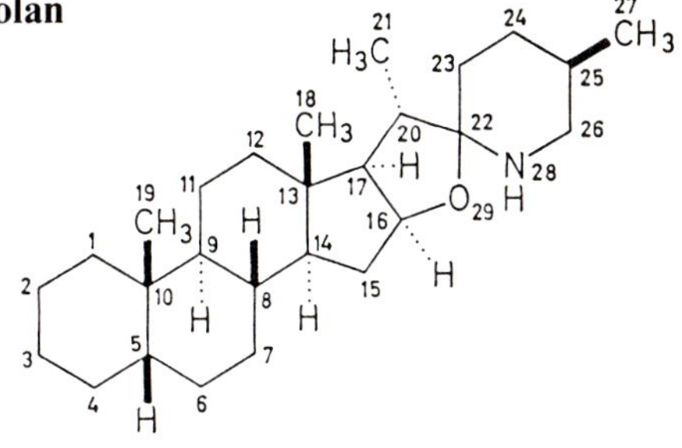

Spirostan **Spirostan**

$C_{27}H_{44}O_2$
Beilstein: 4-19.300 · Syst.-Nr. 2673
CAS-Reg.No.: 6173-22-4
IUPAC: 2S-3.3(47)

–, (25*R*)- = *Isogenine*
–, (25*S*)- = *Neogenine, Sapogenin*

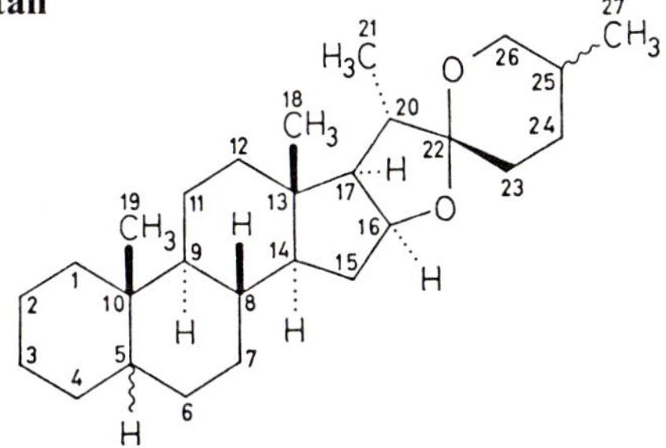

Squalane **Squalan**

= Tetracosane,
2,6,10,15,19,23-hexamethyl-

Squalene **Squalen**

sys[CA]: 2,6,10,14,18,22-Tetracosa-hexaene, 2,6,10,15,19,23-hexamethyl-, (all-*E*)-

$C_{30}H_{50}$
Beilstein: 4-1.1146 · Syst.-Nr. 15
CAS-Reg.No.: 111-02-4

Stachane **Stachan**

= Phyllocladane, 13-methyl-17-nor-

Staphimine **Staphimin**

$C_{41}H_{54}N_2O$
Beilstein: 4-27.7388 · Syst.-Nr. 4501
CAS-Reg.No.: 59588-18-0

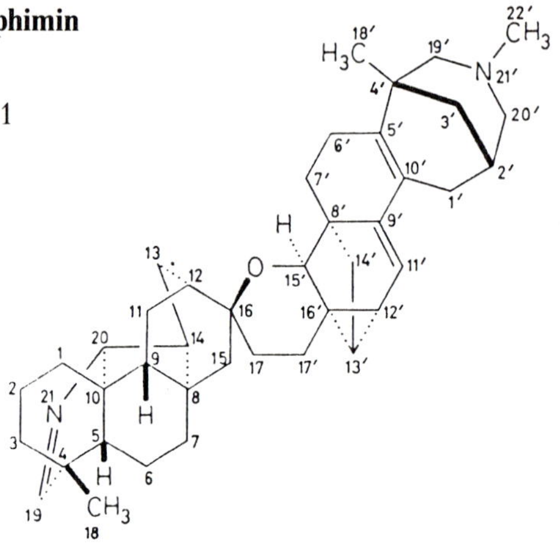

Stearic acid **Stearinsäure**

= Octadecanoic acid

Stercobilin **Stercobilin**

syn: Stercobilin IXα
sys[CA]: 21*H*-Biline-8,12-dipropanoic acid, 3,18-diethyl-1,2,3,4,-5,15,16,17,18,19,22,24-dodecahydro-2,7,13,17-tetramethyl-1,19-dioxo-, (2*R*,3*R*,4*S*,16*S*,-17*R*,18*R*)-

$C_{33}H_{46}N_4O_6$

Beilstein: 4-26.3256 · Syst.-Nr. 4173
CAS-Reg.No.: 34217-90-8
IUPAC: TP-6.4

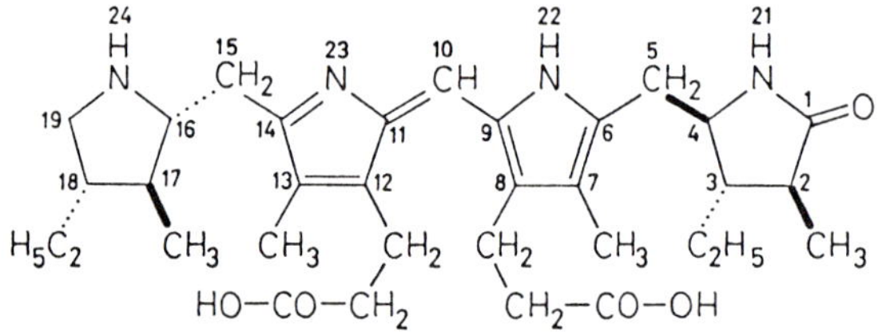

Stigmastane **Stigmastan**

$C_{29}H_{52}$
Beilstein: 3-5.1145 · Syst.-Nr. 472
CAS-Reg.No.: 601-58-1
IUPAC: 2S-2.3(20)

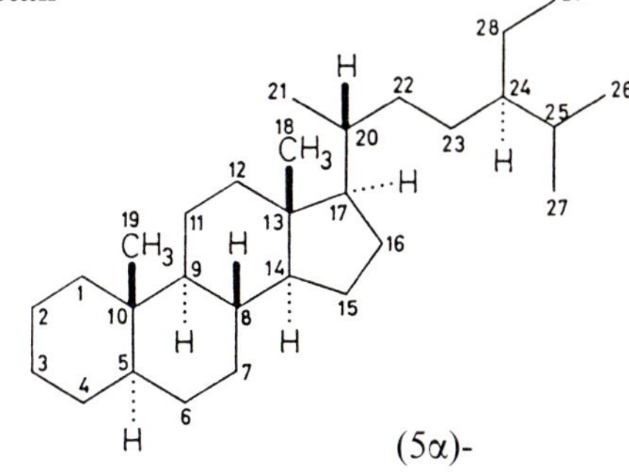

Stilbene **Stilben**

sys[CA]: Benzene,
1,1′-(1,2-ethenediyl)bis-

$C_{14}H_{12}$
Beilstein · Syst.-Nr. 480
CAS-Reg.No.: 588-59-0
IUPAC-Regel: A-61.6

cis-Stilbene
Beilstein: 4-5.2155
CAS-Reg.No.: 645-49-8

trans-Stilbene
Beilstein: 4-5.2156
CAS-Reg.No.: 103-30-0

Streptamine **Streptamin**

sys[B]: Cyclohexan-1,2,3,5-tetrol,
4,6-Diamino-

$C_6H_{14}N_2O_4$
Beilstein: 4-13.2986 · Syst.-Nr. 1871
CAS-Reg.No.: 488-52-8

Streptovaricinoic acid **Streptovaricinsäure**

$C_{39}H_{49}NO_{14}$
Beilstein: 4-27.6713* · Syst.-Nr. 4454
CAS-Reg.No.: 35413-62-8

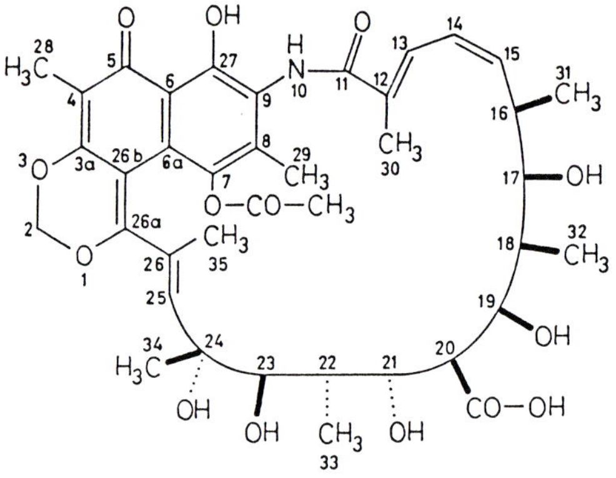

Strophanthidin **Strophanthidin**

= Card-20(22)-enolide,
3β,5,14-trihydroxy-19-oxo-(5β,14β)-

Strychnidine **Strychnidin**

$C_{21}H_{24}N_2O$
Beilstein: 4-27.7229 · Syst.-Nr. 4496
CAS-Reg.No.: 2054-81-1

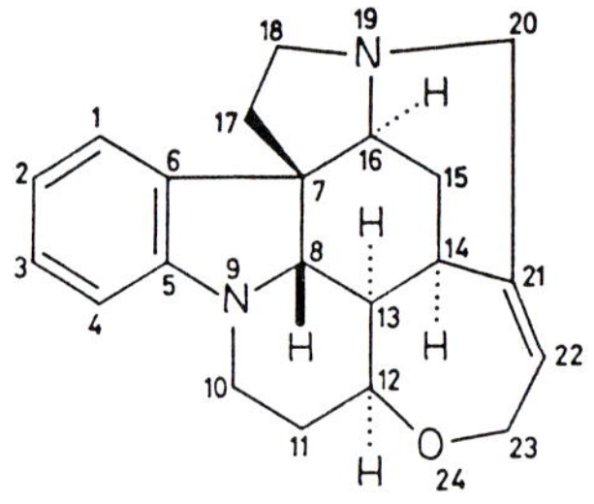

Styphnic acid **Styphninsäure**
sys[CA]: 1,3-Benzenediol, 2,4,6-trinitro-

$C_6H_3N_3O_8$
Beilstein: 4-6.5699 · Syst.-Nr. 554
CAS-Reg.No.: 82-71-3
IUPAC: C-202.2

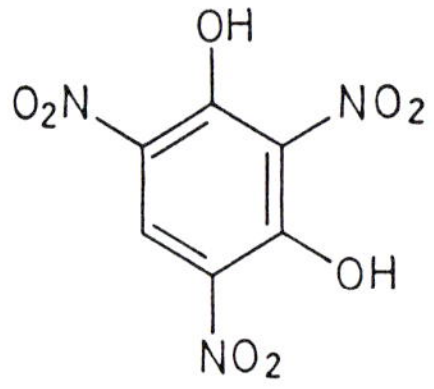

Styrene **Styrol**
sys[CA]: Benzene, ethenyl-

C_8H_8
Beilstein: 4-5.1334 · Syst.-Nr. 473
CAS-Reg.No.: 100-42-5
IUPAC: A-12.1

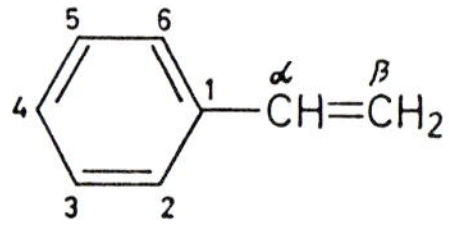

Suberic acid **Korksäure**

= Octanedioic acid

Succinaldehyde **Succinaldehyd**
sys[CA]: Butanedial

$C_4H_6O_2$
Beilstein: 4-1.3642 · Syst.-Nr. 95
CAS-Reg.No.: 638-37-9
IUPAC: C-305.3

$OCH-CH_2-CH_2-CHO$

Succinic acid **Bernsteinsäure**
sys[CA]: Butanedioic acid

$C_4H_6O_4$
Beilstein: 4-2.1908 · Syst.-Nr. 172
CAS-Reg.No.: 110-15-6
IUPAC: C-404.1

$HO-CO-CH_2-CH_2-CO-OH$

–, hydroxy- = Malic acid

Sucrose **Saccharose**

= α-D-Glucopyranoside,
β-D-fructofuranosyl-

Sulfanilic acid **Sulfanilsäure**
sys[CA]: Benzenesulfonic acid,
4-amino-

$C_6H_7NO_3S$
Beilstein: 4-14.2655 · Syst.-Nr. 1923 B_3
CAS-Reg.No.: 121-57-3
IUPAC: C-641.1

H_2N–(4)C₆H₄(1)–SO_2-OH (ring positions 1–6)

Swietenose **Swietenose**

= D-Galactose,
O^6-α-D-galactopyranosyl-

Tacalciol **Tacalciol**

syn: $Tachysterol_3$
sys[CA]: 9,10-Secocholesta-5(10),6,8-trien-3-ol

$C_{27}H_{44}O$
Beilstein: 4-6.4148 · Syst.-Nr. 535
CAS-Reg.No.: 17592-07-3
IUPAC: VD-5

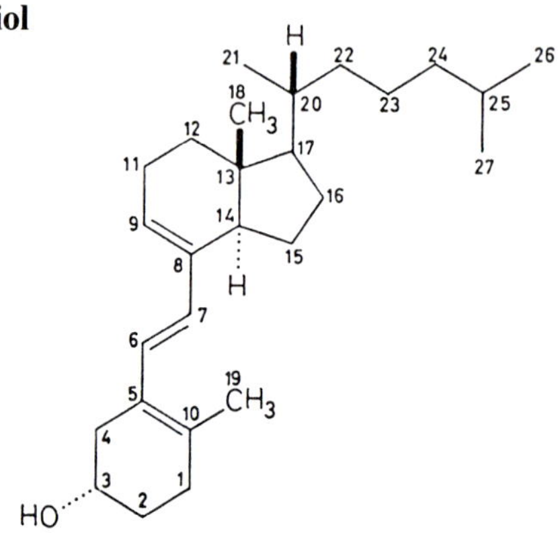

$Tachysterol_3$ **$Tachysterin_3$**

= Tacalciol

Tagatofuranose **Tagatofuranose**

$C_6H_{12}O_6$
Beilstein · Syst.-Nr. 145
IUPAC: Carb-10/18

α-D-Tagatofuranose
CAS-Reg.No.: 36441-92-6
β-D-Tagatofuranose
CAS-Reg.No.: 40461-86-7
α-L-Tagatofuranose
CAS-Reg.No.: 41846-99-5
β-L-Tagatofuranose
CAS-Reg.No.: 41846-98-4

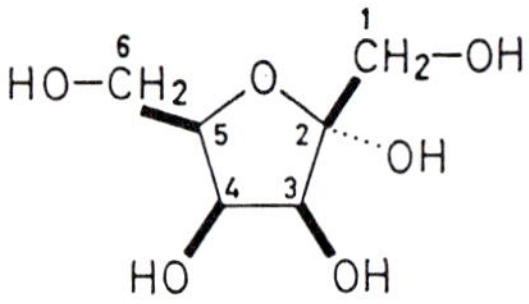

α-D-

Tagatopyranose **Tagatopyranose**

$C_6H_{12}O_6$
Beilstein · Syst.-Nr. 145
IUPAC: Carb-10/18

α-D-Tagatopyranose
Beilstein: 4-1.4414
CAS-Reg.No.: 512-20-9
β-D-Tagatopyranose
CAS-Reg.No.: 20197-42-6

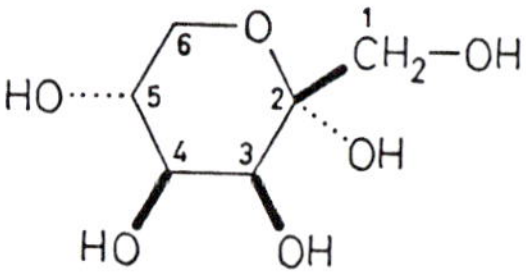

α-D-

Tagatose **Tagatose**

$C_6H_{12}O_6$
Beilstein · Syst.-Nr. 145
CAS-Reg.No.: 17598-81-1
IUPAC: Carb-10

D-Tagatose
Beilstein: 4-1.4414
CAS-Reg.No.: 87-81-0
L-Tagatose
Beilstein: 4-1.4415
CAS-Reg.No.: 17598-82-2
DL-Tagatose
Beilstein: 4-1.4415
CAS-Reg.No.: 73952-11-1

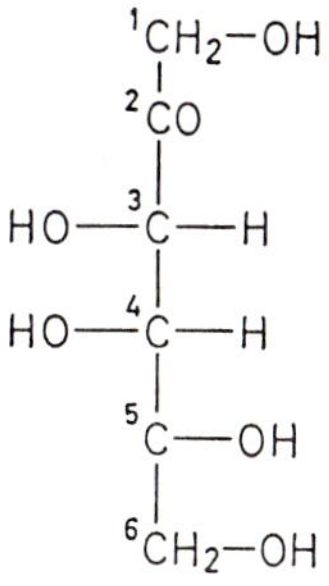

D-

Talbotine **Talbotin**

$C_{21}H_{24}N_2O_4$
Beilstein · Syst.-Nr. 3702
CAS-Reg.No.: 30809-15-5

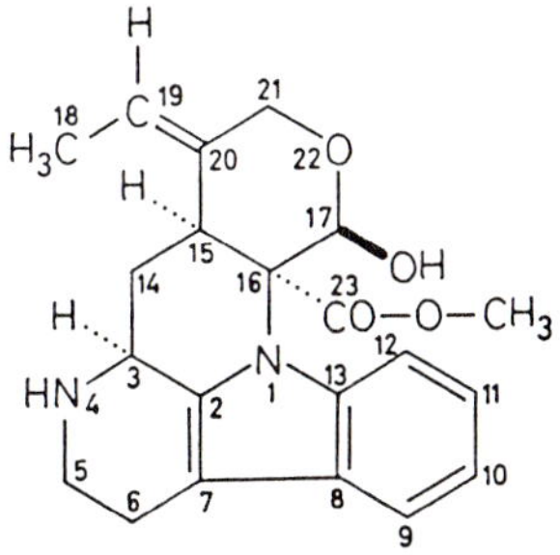

Talofuranose **Talofuranose**

$C_6H_{12}O_6$
Beilstein · Syst.-Nr. 144
IUPAC: Carb-5/18

α-D-Talofuranose
Beilstein: 4-1.4345
CAS-Reg.No.: 51076-04-1
β-D-Talofuranose
Beilstein: 4-1.4345
CAS-Reg.No.: 41847-63-6
α-L-Talofuranose
CAS-Reg.No.: 36978-43-5
β-L-Talofuranose
CAS-Reg.No.: 40461-83-4

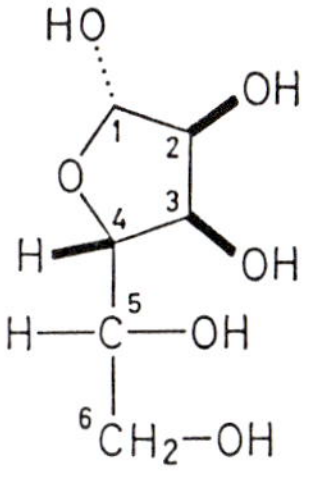

α-D-

Talopyranose **Talopyranose**

$C_6H_{12}O_6$
Beilstein · Syst.-Nr. 144
IUPAC: Carb-5/18

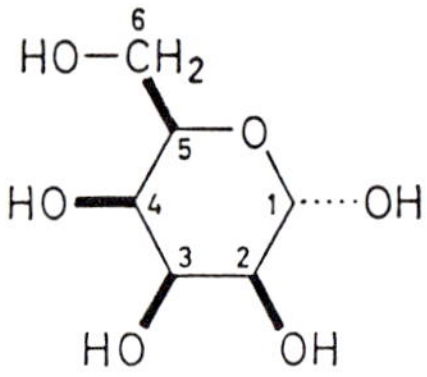

α-D-Talopyranose
Beilstein: 4-1.4345
CAS-Reg.No.: 7282-81-7
β-D-Talopyranose
Beilstein: 4-1.4345
CAS-Reg.No.: 7283-11-6
α-L-Talopyranose
CAS-Reg.No.: 12773-32-9
β-L-Talopyranose
CAS-Reg.No.: 54808-89-8

α-D-

Talose **Talose**

$C_6H_{12}O_6$
Beilstein · Syst.-Nr. 144
CAS-Reg.No.: 30077-17-9
IUPAC: Carb-5

–, O^3-methyl-6-deoxy- = *Acovenose*

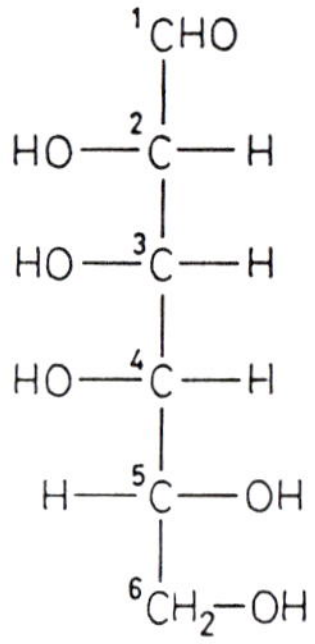

D-Talose
Beilstein: 4-1.4344
CAS-Reg.No.: 2595-98-4
L-Talose
Beilstein: 4-1.4346
CAS-Reg.No.: 23567-25-1

D-

Taraxastane **Taraxastan**
$C_{30}H_{52}$
Beilstein: 3-5.1340 · Syst.-Nr. 473

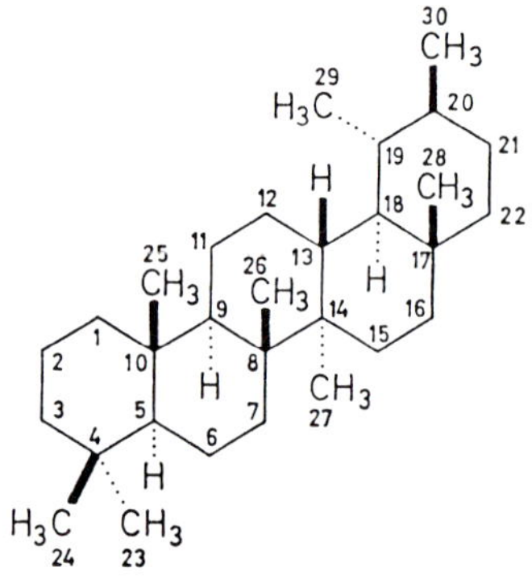

Taraxerane **Taraxeran**

sys[CA]: *D*-Friedooleanane, (14α)-

$C_{30}H_{52}$
Beilstein: 3-5.1342 · Syst.-Nr. 473
CAS-Reg.No.: 39865-11-7

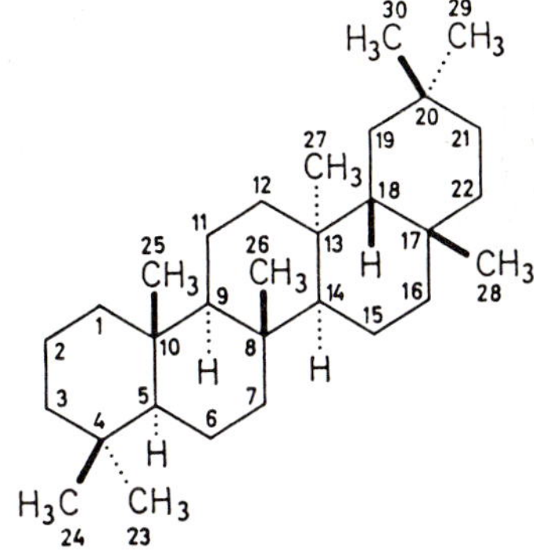

Tartaric acid **Weinsäure**

sys[B]: Bernsteinsäure, 2,3-Dihydroxy-
[CA]: Butanedioic acid, 2,3-dihydroxy-

$C_4H_6O_6$
Beilstein · Syst.-Nr. 250
CAS-Reg.No.: 526-83-0
IUPAC: C-411.1

HO – CO – CH(OH) – CH(OH) – CO – OH

D_g-Tartaric acid
Beilstein: 4-3.1229
CAS-Reg.No.: 147-71-7

L_g-Tartaric acid
Beilstein: 4-3.1219
CAS-Reg.No.: 87-69-4

DL-Tartaric acid
Beilstein: 4-3.1229
CAS-Reg.No.: 133-37-9

meso-Tartaric acid
Beilstein: 4-3.1218
CAS-Reg.No.: 147-73-9

Tartronic acid **Tartronsäure**

sys[B]: Malonsäure, Hydroxy-
[CA]: Propanedioic acid, Hydroxy-

$C_3H_4O_5$
Beilstein: 4-3.1120 · Syst.-Nr. 239
CAS-Reg.No.: 80-69-3
IUPAC: C-411.1

HO – CO – CH(OH) – CO – OH

Taurine **Taurin**

sys[B]: Äthansulfonsäure, 2-Amino-
[CA]: Ethanesulfonic acid, 2-amino-

$C_2H_7NO_3S$
Beilstein: 4-4.3289 · Syst.-Nr. 379
CAS-Reg.No.: 107-35-7
IUPAC: C-641.1

$H_2N-CH_2-CH_2-SO_2-OH$

Taxane **Taxan**

sys[CA]: 6,10-Methanobenzocyclodecene, tetradecahydro-4,9,12a,13,13-pentamethyl-, [4*R*-(4α,4aβ,6α,9α,10α,12aα)]-

$C_{20}H_{36}$
Beilstein: 3-9.2390* · Syst.-Nr. 461
CAS-Reg.No.: 1605-68-1
IUPAC: F-(22)

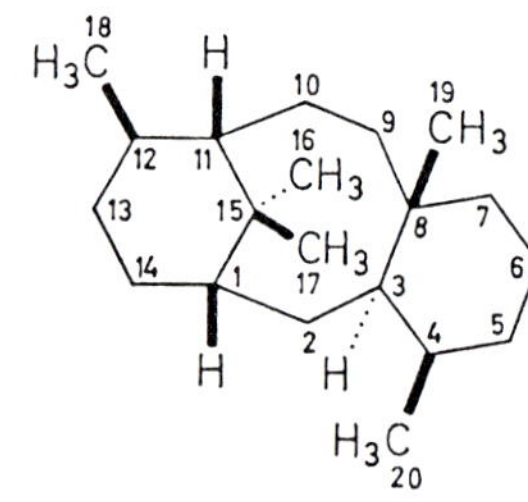

Tazettine **Tazettin**

$C_{18}H_{21}NO_5$
Beilstein: 4-27.6643 · Syst.-Nr. 4447
CAS-Reg.No.: 507-79-9

Telecinobufagin **Telecinobufagin**

= Bufa-20,22-dienolide, 3β,5,14-trihydroxy-(5β,14β)-

Terephthalic acid **Terephthalsäure**

sys[CA]: 1,4-Benzenedicarboxylic acid

$C_8H_6O_4$
Beilstein: 4-9.3301 · Syst.-Nr. 978
CAS-Reg.No.: 100-21-0
IUPAC: C-404.1

Testane **Testan**

= Androstane, (5β)-

Testosterone **Testosteron**

= Androst-4-en-3-one, 17β-hydroxy-

Tetrabenzoporphyrin **Tetrabenzoporphyrin**

$C_{36}H_{22}N_4$
Beilstein: 4-26.1954 · Syst.-Nr. 4034
CAS-Reg.No.: 4466-64-2
IUPAC: TP-1.3

–, 5,10,15,20-tetraaza- = Phthalocyanine

Tetrahedrane **Tetrahedran**
sys[CA]: Tricyclo[1.1.0.0^{2,4}]butane

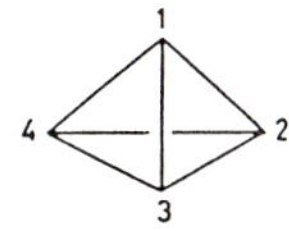

C_4H_4
Beilstein · Syst.-Nr. 455
CAS-Reg.No.: 157-39-1

Thalicberan **Thalicberan**

$C_{32}H_{30}N_2O_2$
Beilstein: 4-27.8739* · Syst.-Nr. 4635
CAS-Reg.No.: 36573-86-1
IUPAC: F-2

Thalidasan **Thalidasan**

$C_{32}H_{30}N_2O_2$
Beilstein · Syst.-Nr. 4635
CAS-Reg.No.: 36573-88-3
IUPAC: F-2

Thalman **Thalman**

$C_{32}H_{30}N_2O_2$
Beilstein: 4-27.8740* · Syst.-Nr. 4635
CAS-Reg.No.: 36573-87-2
IUPAC: F-2

Thebenidine **Thebenidin**

sys[B]: Benzo[*lmn*]phenanthridin

$C_{15}H_9N$
Beilstein: 4-20.4232 · Syst.-Nr. 3090
CAS-Reg.No.: 194-03-6

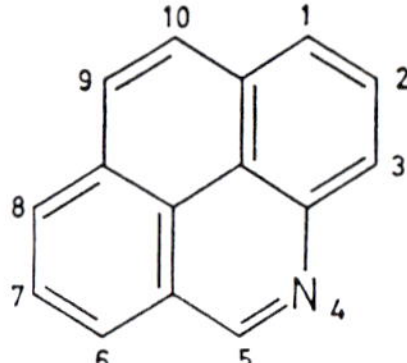

Thevetose **Thevetose**

= Glucose, O^3-methyl-6-deoxy-

Thianaphthene **Thianaphthen**

= Benzo[*b*]thiophene

Thianthrene **Thianthren**

$C_{12}H_8S_2$
Beilstein: 4-19.347 · Syst.-Nr. 2676
CAS-Reg.No.: 92-85-3
IUPAC: B-2.11(4)

Thiobiuret **Thiobiuret**

sys[CA]: Thioimidodicarbonic diamide

$C_2H_5N_3OS$, formula A [IUPAC-numbering and formula B [CA-numbering]
Beilstein: 3-3.305 · Syst.-Nr. 216
CAS-Reg.No.: 23228-74-2
IUPAC: C-975.1

$\overset{5}{H_2N}-\overset{4}{CO}-\overset{3}{NH}-\overset{2}{CS}-\overset{1}{NH_2}$

A

$\overset{N}{H_2N}-\overset{1}{CO}-\overset{2}{NH}-\overset{3}{CS}-\overset{N'}{NH_2}$

B

Thiophanthrene **Thiophanthren**

= Naphtho[2,3-*b*]thiophene

Thiophene **Thiophen**

C_4H_4S
Beilstein: 5-17/1.297 · Syst.-Nr. 2364
CAS-Reg.No.: 110-02-1
IUPAC: B-2.11(1)

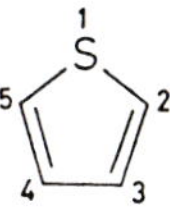

Thiuram disulfide **Thiuramdisulfid**

sys[CA]: Thioperoxydicarbonic diamide

$C_2H_4N_2S_4$
Beilstein: 3-3.355 · Syst.-Nr. 218
CAS-Reg.No.: 504-90-5
IUPAC: C-661.6

$H_2N-CS-S-S-CS-NH_2$

Thiuram monosulfide **Thiuramsulfid**

sys[CA]: Thiodicarbonic diamide

$C_2H_4N_2S_3$
Beilstein: 3-3.355 · Syst. Nr. 218
CAS-Reg.No.: 503-82-2
IUPAC: C-661.6

$H_2N-CS-S-CS-NH_2$

Threonine **Threonin**

$C_4H_9NO_3$
Beilstein · Syst.-Nr. 376
CAS-Reg.No.: 36676-50-3
IUPAC: 3AA-1

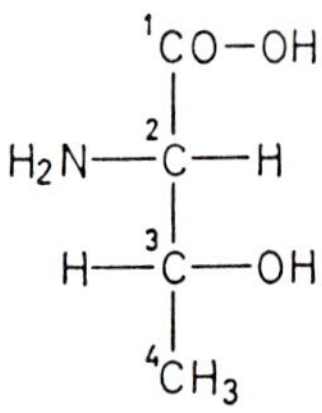

D_s-Threonine
Beilstein: 4-4.3171
CAS-Reg.No.: 632-20-2
L_s-Threonine
Beilstein: 4-4.3171
CAS-Reg.No.: 72-19-5
DL-Threonine
Beilstein: 4-4.3172
CAS-Reg.No.: 80-68-2

^{1}CO—OH
H_2N—^{2}C—H
H—^{3}C—OH
$^4CH_3$

L_s-

Threose **Threose**

sys[CA]: Butanal, 2,3,4-trihydroxy-, (*R**,*S**)-

$C_4H_8O_4$
Beilstein · Syst.-Nr. 124
CAS-Reg.No.: 29884-64-8
IUPAC: Carb-5

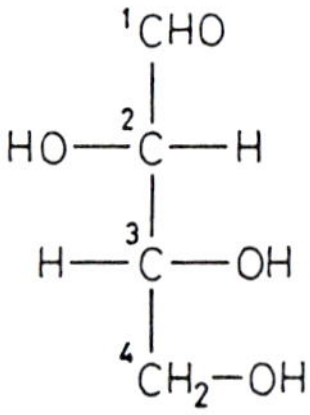

D-Threose
Beilstein: 4-1.4173
CAS-Reg.No.: 95-43-2
L-Threose
Beilstein: 4-1.4173
CAS-Reg.No.: 95-44-3
DL-Threose
Beilstein: 4-1.4173
CAS-Reg.No.: 38982-45-5

^{1}CHO
HO—^{2}C—H
H—^{3}C—OH
$^4CH_2$—OH

D-

Thromboxane **Thromboxan**

$C_{20}H_{40}O$
Beilstein · Syst.-Nr. 2362
CAS-Reg.No.: 66719-58-2
IUPAC: F-2

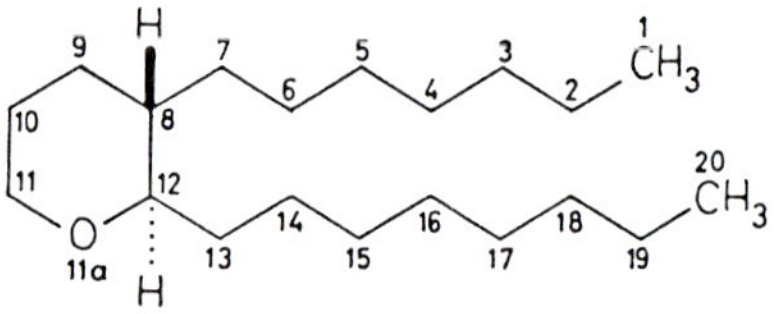

Thujane **Thujan**

sys[CA]: Bicyclo[3.1.0]hexan, 4-methyl-1-(1-methylethyl)-

$C_{10}H_{18}$
Beilstein: 4-5.317 · Syst.-Nr. 453
CAS-Reg.No.: 471-12-5
IUPAC: A-72.1

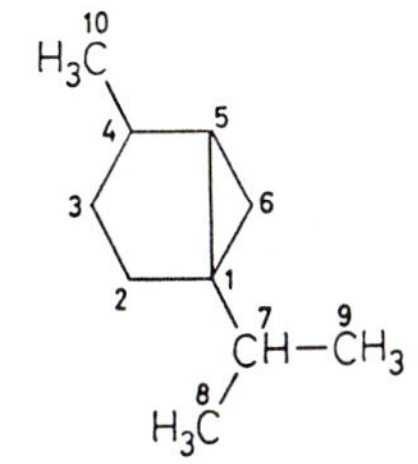

Thymidine **Thymidin**

$C_{10}H_{14}N_2O_5$
Beilstein: 4-24.1297 · Syst.-Nr. 3588
CAS-Reg.No.: 50-89-5

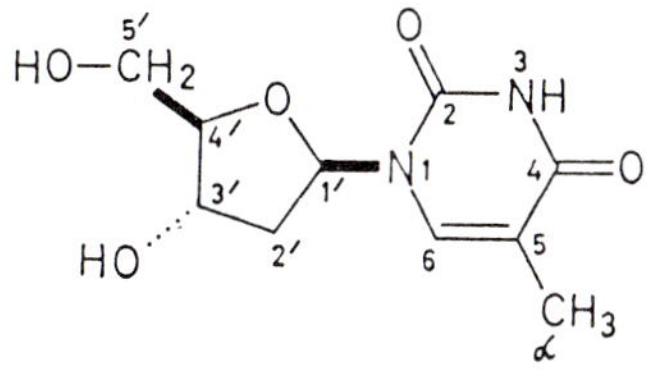

Thymine **Thymin**

sys[B]: 1*H*-Pyrimidin-2,4-dion, 5-Methyl-
[CA]: 2,4(1*H*,3*H*)-Pyrimidinedione, 5-methyl-

$C_5H_6N_2O_2$
Beilstein: 4-24.1292 · Syst.-Nr. 3588
CAS-Reg.No.: 65-71-4

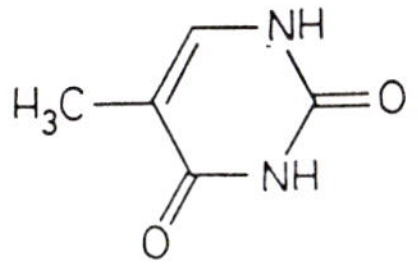

Thyminose **Thyminose**

= D-*erythro*-2-Deoxy-pentose

Thymol **Thymol**

sys[B]: Phenol, 2-Isopropyl-5-methyl-
[CA]: Phenol, 5-methyl-2-(1-methylethyl)-

$C_{10}H_{14}O$
Beilstein: 4-6.3334 · Syst.-Nr. 532
CAS-Reg.No.: 89-83-8
IUPAC: C-202.2

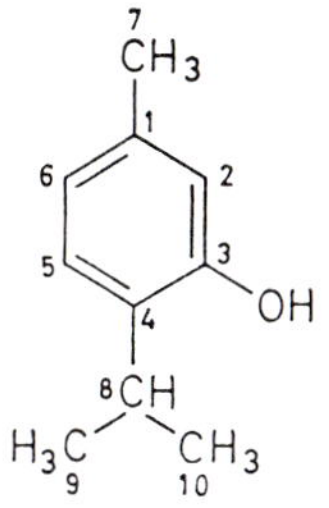

Thyronine **Thyronin**

sys[CA]: Tyrosine, *O*-(4-hydroxyphenyl)-

$C_{15}H_{15}NO_4$
Beilstein · Syst.-Nr. 1911
CAS-Reg.No.: 101-66-6
IUPAC: C-421.1; 3AA-Appendix

L-Thyronine
Beilstein: 3-14.1508
CAS-Reg.No.: 1596-67-4

DL-Thyronine
Beilstein: 4-14.2269
CAS-Reg.No.: 1034-10-2

Thyroxine **Thyroxin**

= Thyronine, 3,5,3′,5′-tetrajodo-

Tigliane **Tiglian**

sys[CA]: 1*H*-Cyclopropa[3,4]benz[1,2-*e*]azulene, tetradecahydro-1,1,3,6,8-pentamethyl-, [1a*S*-(1aα,1bβ,3β,4aβ,6β,7aα,7bα,8α,9aα)]-

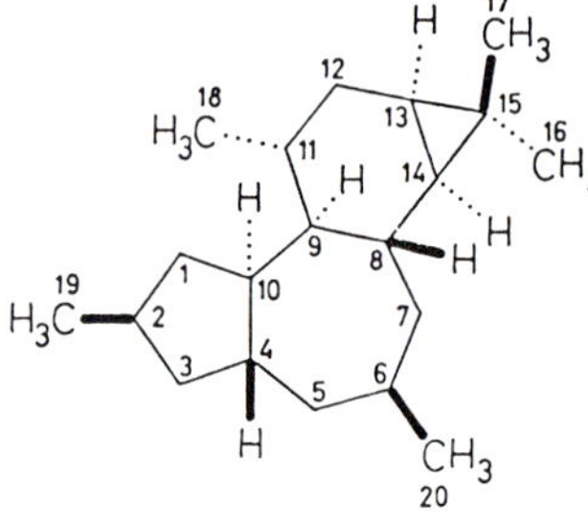

$C_{20}H_{34}$
Beilstein: 3-8.4217* · Syst.-Nr. 472
CAS-Reg.No.: 67707-87-3
IUPAC: F-2

Tiglic acid **Tiglinsäure**

= *trans*-Crotonic acid, 2-methyl-

Tirucallane **Tirucallan**

sys[CA]: Lanostane, (13α,14β,17α,20*S*)-

$C_{30}H_{54}$
Beilstein: 3-8.1222* · Syst.-Nr. 472
IUPAC: 2S-2.3(28)

Tocol **Tocol**

sys[B]: Chroman-6-ol, 2-Methyl-2-[4,8,12-trimethyl-tridecyl]-

[CA]: 2*H*-1-Benzopyran-6-ol, 3,4-dihydro-2-methyl-2-(4,8,12-trimethyltridecyl)-

$C_{26}H_{44}O_2$
Beilstein: 5-17/4.129 · Syst.-Nr. 2384
CAS-Reg.No.: 119-98-2
IUPAC: Toc-1.2

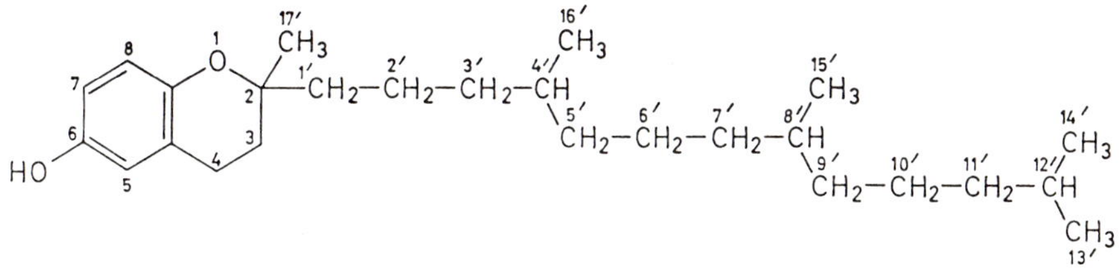

Toluene **Toluol**

sys[CA]: Benzene, methyl-

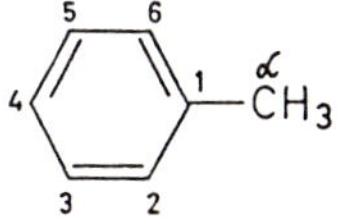

C_7H_8
Beilstein: 4-5.766 · Syst.-Nr. 466
CAS-Reg.No.: 108-88-3
IUPAC: A-12.1

Tomatanine **Tomatanin**

= Spirosolane, (5α)-

Tomatidane **Tomatidan**

= Spirosolane, (5α)-

Totarane **Totaran**

= Podocarpane, 14-isopropyl-

Toxiferin I **Toxiferin-I**

$[C_{40}H_{46}N_4O_2]^{2+}$
Beilstein: 4-26.2041 · Syst.-Nr. 4082
CAS-Reg.No.: 6888-23-9

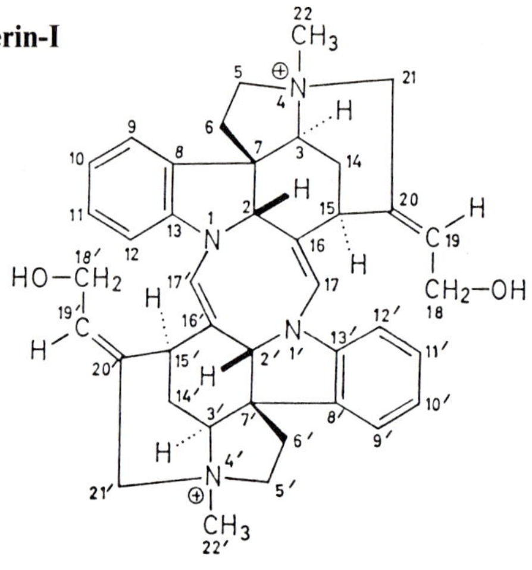

Trehalose **Trehalose**

= D-Glucopyranoside, D-glucopyranosyl-

Tretinoin **Tretinoin**

= Retinoic acid

Trichothecane **Trichothecan**

$C_{15}H_{26}O$
Beilstein: 4-17.2337* · Syst.-Nr. 2364
CAS-Reg.No.: 24706-08-9
IUPAC: F-2

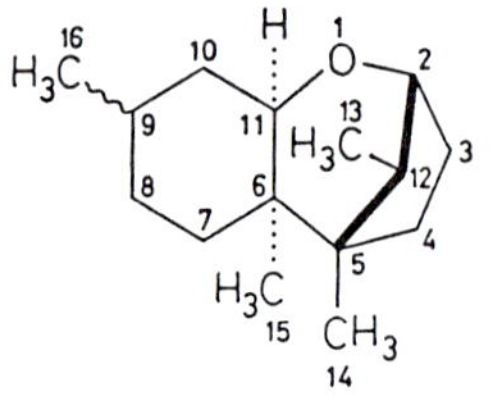

Trinaphthylene **Trinaphthylen**

$C_{30}H_{18}$
Beilstein: 3-5.2719 · Syst.-Nr. 497
CAS-Reg.No.: 196-62-3

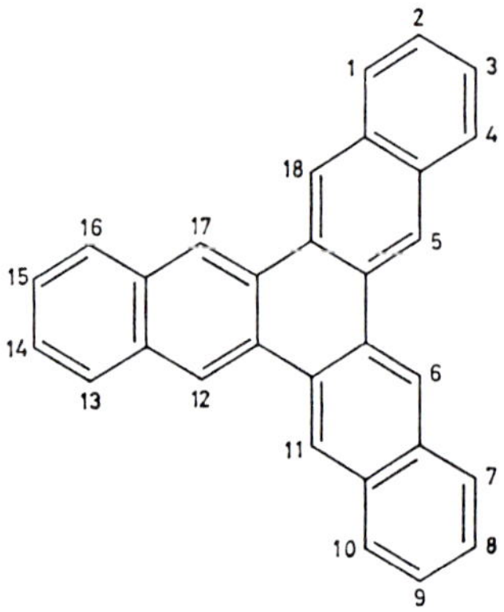

Trindene **Trinden**

$C_{15}H_{10}$
Beilstein · Syst.-Nr. 486

1*H*-Trindene
CAS-Reg.No.: 228-30-8
3a*H*-Trindene
CAS-Reg.No.: 228-29-5

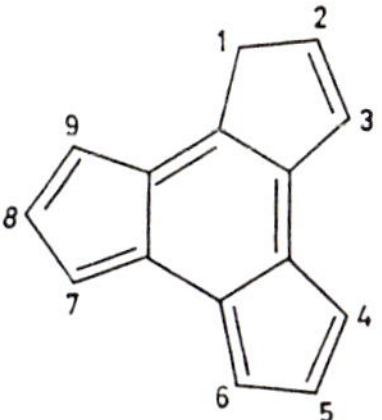

Triphenylene **Triphenylen**

$C_{18}H_{12}$
Beilstein: 4-5.2556 · Syst.-Nr. 488
CAS-Reg.No.: 217-59-4
IUPAC: A-21.1(17)

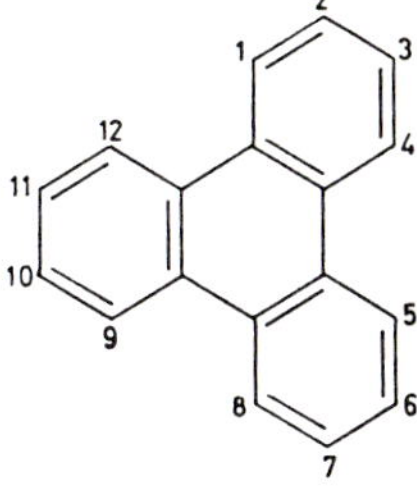

Tripyrrin **Tripyrrin**

sys[CA]: 1*H*-Pyrrole, 2-[[5-(2*H*-pyrrol-2-ylidenemethyl)-2*H*-pyrrol-2-ylidene]methyl]-

$C_{14}H_{11}N_3$
Beilstein: 4-26.954 · Syst.-Nr. 3813
CAS-Reg.No.: 58282-09-0
IUPAC: TP-7.1

Triuret **Triuret**

sys[CA]: Diimidotricarbonic diamide

$C_3H_6N_4O_3$

$H_2N-CO-NH-CO-NH-CO-NH_2$

Beilstein: 3-3.142 · Syst.-Nr. 205 A
CAS-Reg.No.: 556-99-0
IUPAC: C-975.1

Tropane **Tropan**

sys[CA]: 8-Azabicyclo[3.2.1]octane, 8-methyl-

$C_8H_{15}N$
Beilstein: 4-20.1963 · Syst.-Nr. 3047
CAS-Reg.No.: 529-17-9

Tropic acid **Tropasäure**

sys[B]: Propionsäure, 3-Hydroxy-2-phenyl-
[CA]: Benzeneacetic acid, α-(hydroxymethyl)-

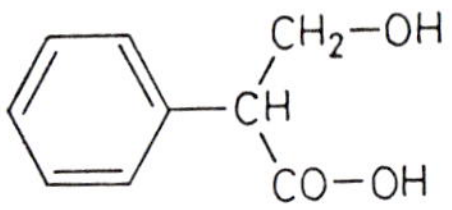

$C_9H_{10}O_3$
Beilstein · Syst.-Nr. 1073
CAS-Reg.No.: 529-64-6
IUPAC: C-411.1

(*R*)-Tropic acid
Beilstein: 4-10.664
CAS-Reg.No.: 17126-67-9
(*S*)-Tropic acid
Beilstein: 4-10.664
CAS-Reg.No.: 16202-15-6
(±)-Tropic acid
Beilstein: 4-10.664
CAS-Reg.No.: 552-63-6

Tryptophane **Tryptophan**

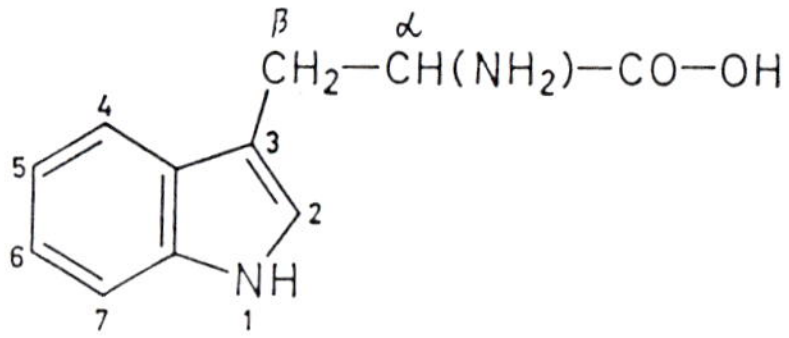

$C_{11}H_{12}N_2O_2$
Beilstein · Syst.-Nr. 3436
CAS-Reg.No.: 6912-86-3
IUPAC: 3AA-2.3.3

D-Tryptophane
Beilstein: 4-22.6765
CAS-Reg.No.: 153-94-6
L-Tryptophane
Beilstein: 4-22.6765
CAS-Reg.No.: 73-22-3
DL-Tryptophane
Beilstein: 4-22.6768
CAS-Reg.No.: 54-12-6

Tubocuraran **Tubocuraran**

$C_{32}H_{30}N_2O_2$
Beilstein: 4-19.4282* · Syst.-Nr. 4635
CAS-Reg.No.: 34561-13-2
IUPAC: F-2

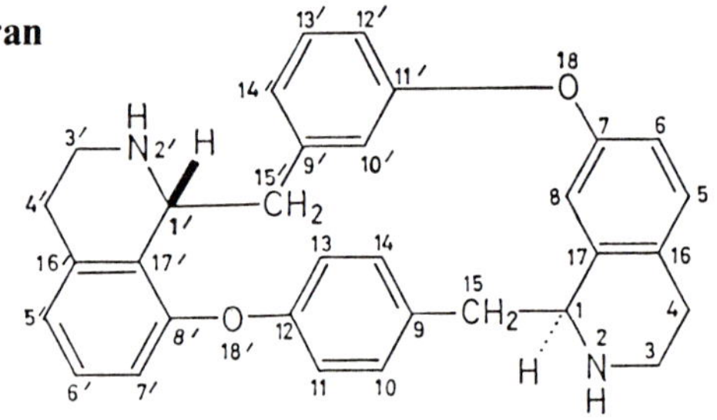

Tubulosan **Tubulosan**

$C_{27}H_{33}N_3$
Beilstein · Syst.-Nr. 3815
CAS-Reg.No.: 36104-77-5
IUPAC: F-2

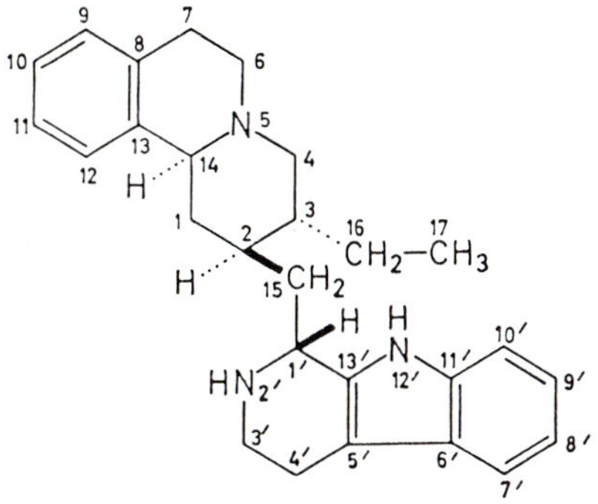

Turanose **Turanose**

= D-Fructose, O^3-α-D-glucopyranosyl-

Tyrosine **Tyrosin**

$C_9H_{11}NO_3$
Beilstein · Syst.-Nr. 1911
CAS-Reg.No.: 55520-40-6
IUPAC: 3AA-2.2.3

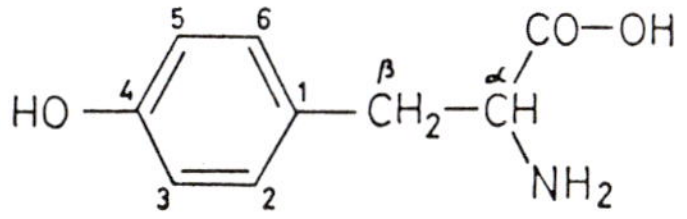

D-Tyrosine
Beilstein: 4-14.2264
CAS-Reg.No.: 556-02-5
L-Tyrosine
Beilstein: 4-14.2264
CAS-Reg.No.: 60-18-4
DL-Tyrosine
Beilstein: 4-14.2266
CAS-Reg.No.: 556-40-6

Tyvelose **Tyvelose**

= D-*arabino*-3,6-Dideoxy-hexose

Uracil **Uracil**

sys[CA]: 2,4(1*H*,3*H*)-Pyrimidinedione

$C_4H_4N_2O_2$
Beilstein: 4-24.1193 · Syst.-Nr. 3588
CAS-Reg.No.: 66-22-8

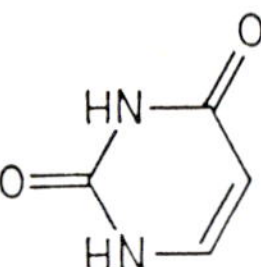

Urea **Harnstoff**

CH_4N_2O
Beilstein: 4-3.94 · Syst.-Nr. 205 A
CAS-Reg.No.: 57-13-6
IUPAC: C-971.1

$H_2N-CO-NH_2$

Uridine **Uridin**

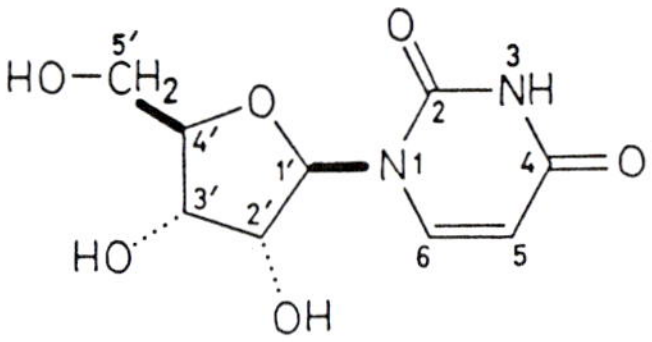

$C_9H_{12}N_2O_6$
Beilstein: 4-24.1202 · Syst.-Nr. 3588
CAS-Reg.No.: 58-96-8

Urobilin **Urobilin**

syn: Urobilin IXα

sys[CA]: 21*H*-Biline-8,12-dipropanoic acid, 3,18-diethyl-1,4,5,15,-16,19,22,24-octahydro-2,7,-13,17-tetramethyl-1,19-dioxo-

$C_{33}H_{42}N_4O_6$
Beilstein: 4-26.3258 · Syst.-Nr. 4173
CAS-Reg.No.: 1856-98-0
IUPAC: TP-6.4

–, 10,23-dihydro- = Urobilinogen
–, 2,3,17,18-tetrahydro- = Stercobilin

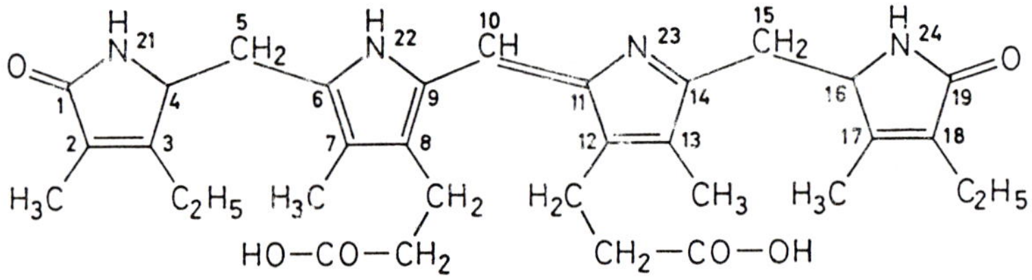

Urobilinogen **Urobilinogen**

syn: Urobilinogen IXα,
Mesobilirubinogen IXα

sys[CA]: 21*H*-Biline-8,12-dipropanoic acid, 2,17-diethyl-1,4,5,10,-15,16,19,22,24-decahydro-3,7,13,18-tetramethyl-1,19-dioxo-

$C_{33}H_{44}N_4O_6$
Beilstein: 4-26.3257 · Syst.-Nr. 4173
CAS-Reg.No.: 14684-37-8
IUPAC: TP-6.4

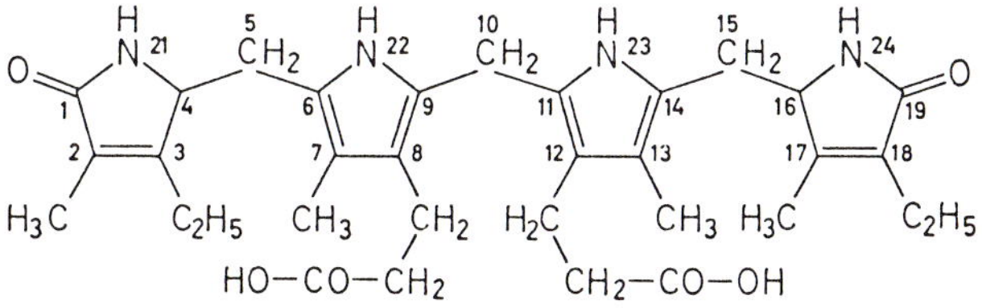

Uroporphyrin I **Uroporphyrin-I**

sys[CA]: 21*H*,23*H*-Porphine-2,7,12,17-tetrapropanoic acid, 3,8,13,18-tetrakis-(carboxymethyl)-

$C_{40}H_{38}N_4O_{16}$
Beilstein: 4-26.3132 · Syst.-Nr. 4173
CAS-Reg.No.: 607-14-7
IUPAC: TP-2.1

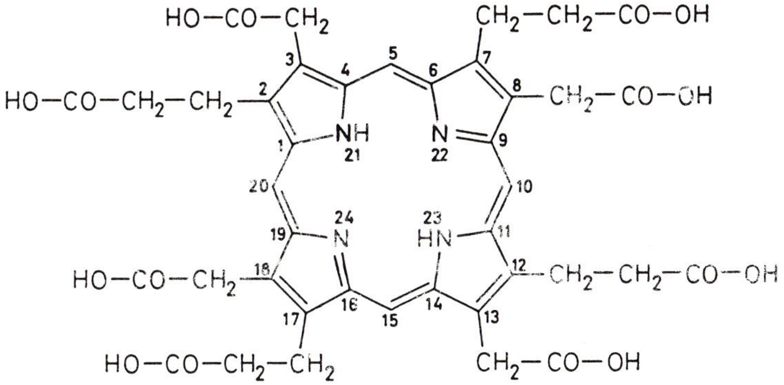

Ursane **Ursan**

$C_{30}H_{52}$
Beilstein: 4-5.1522 · Syst.-Nr. 473
CAS-Reg.No.: 464-93-7
IUPAC: F-(51)

–, (18α,19β*H*,20α*H*)- = Taraxastane

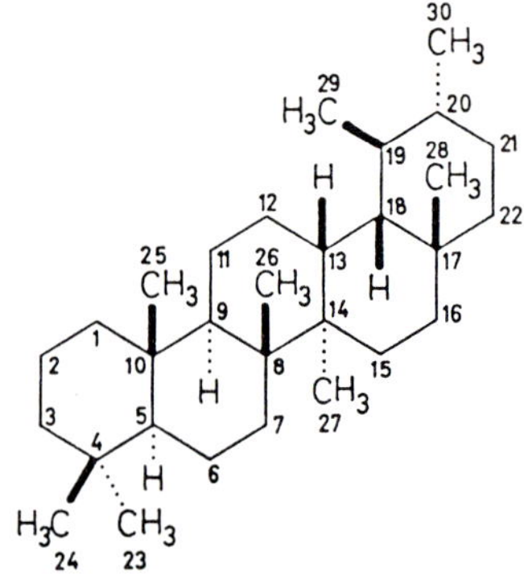

Valeric acid **Valeriansäure**
sys[CA]: Pentanoic acid

$C_5H_{10}O_2$
Beilstein: 4-2.868 · Syst.-Nr. 162
CAS-Reg.No.: 109-52-4
IUPAC: C-404.1

$H_3C-[CH_2]_3-CO-OH$

Valine **Valin**

$C_5H_{11}NO_2$
Beilstein: 4-4.2659 · Syst.-Nr. 367
CAS-Reg.No.: 7004-03-7
IUPAC: 3AA-2.2.1

H_3C (4)
\\ (3) (2) (1)
$CH-CH(NH_2)-CO-OH$
/
H_3C (4′)

D-Valine
Beilstein: 4-4.2659
CAS-Reg.No.: 640-68-6

L-Valine
Beilstein: 4-4.2659
CAS-Reg.No.: 72-18-4

DL-Valine
Beilstein: 4-4.2660
CAS-Reg.No.: 516-06-3

Vanillic acid **Vanillinsäure**

sys[B]: Benzoesäure, 4-Hydroxy-3-methoxy-
[CA]: Benzoic acid, 4-hydroxy-3-methoxy-

$C_8H_8O_4$
Beilstein: 4-10.1459 · Syst.-Nr. 1105
CAS-Reg.No.: 121-34-6
IUPAC: C-411.1

Vanillin **Vanillin**

sys[CA]: Benzaldehyde, 4-hydroxy-3-methoxy-

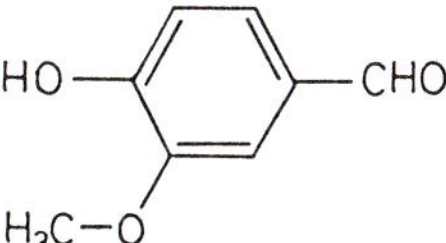

$C_8H_8O_3$
Beilstein: 4-8.1763 · Syst.-Nr. 773 A
CAS-Reg.No.: 121-33-5
IUPAC: C-305.2

Veatchane **Veatchan**

$C_{19}H_{31}N$
Beilstein: 4-20.3329* · Syst.-Nr. 3065
CAS-Reg.No.: 41904-80-7
IUPAC: F-2

Veratraman **Veratraman**

sys[B]: Veratra-5,12-dien

$C_{27}H_{43}N$
Beilstein · Syst.-Nr. 3082
CAS-Reg.No.: 39608-81-6

–, 5,6,12,13-tetrahydro- = Veratran

Veratran **Veratran**

sys[CA]: Veratraman, 5,6,12,13-tetrahydro-

$C_{27}H_{47}N$
Beilstein: 4-20.3713* · Syst.-Nr. 3065
IUPAC: F-2

–, 17,23-Epoxy- = Jervanine

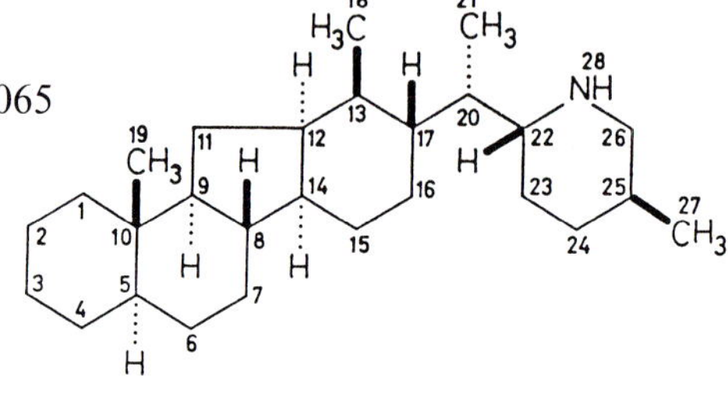

Veratranine **Veratranin**

–, (5α,17β*H*,22*R*,25*S*)- = Veratran

Veratric acid **Veratrumsäure**

sys[B]: Benzoesäure, 3,4-Dimethoxy-
[CA]: Benzoic acid, 3,4-dimethoxy-

$C_9H_{10}O_4$
Beilstein: 4-10.1460 · Syst.-Nr. 1105
CAS-Reg.No.: 93-07-2
IUPAC: C-411.1

Veratrole **Veratrol**

sys[CA]: Benzene, 1,2-dimethoxy-

$C_8H_{10}O_2$
Beilstein: 4-6.5564 · Syst.-Nr. 553
CAS-Reg.No.: 91-16-7
IUPAC: C-214.1

Vicianose **Vicianose**

= D-Glucose, O^6-α-L-arabinopyranosyl-

Violanthrene **Violanthren**

sys[CA]: Anthra[9,1,2-*cde*]benzo[*rst*]-pentaphene, 5,10-dihydro-

$C_{34}H_{20}$
Beilstein: 1-5.392 · Syst.-Nr. 497
CAS-Reg.No.: 81-31-2
IUPAC: A-23.1

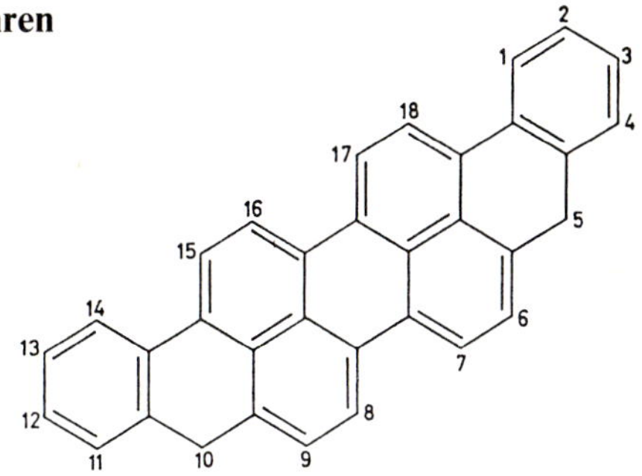

Vitamin A **Vitamin-A**

= Retinol

Vitamin B_2 **Vitamin-B_2**

= Riboflavin

Vitamin B_6 **Vitamin-B_6**

= Pyridoxine

Vitamin C **Vitamin-C**

= L-Ascorbic acid

Vitamin D_2 **Vitamin-D_2**

= Ercalciol

Vitamin A acid **Vitamin-A-säure**

= Retinoic acid

Vitamin A alcohol **Vitamin-A-alkohol**

= Retinol

Vitamin A aldehyde **Vitamin-A-aldehyd**

= Retinal

Vobasan **Vobasan**

$C_{20}H_{26}N_2$
Beilstein: 4-25.1705 · Syst.-Nr. 3486
CAS-Reg.No.: 34486-92-5
IUPAC: F-2

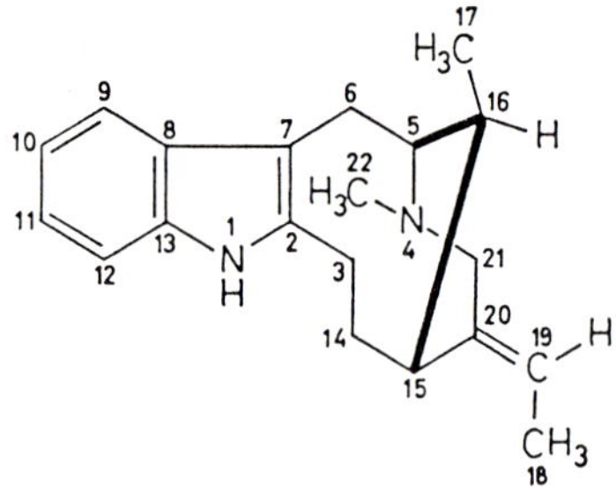

Vobtusine **Vobtusin**

$C_{43}H_{50}N_4O_6$
Beilstein: 4-27.9686 · Syst.-Nr. 4710
CAS-Reg.No.: 19772-79-3

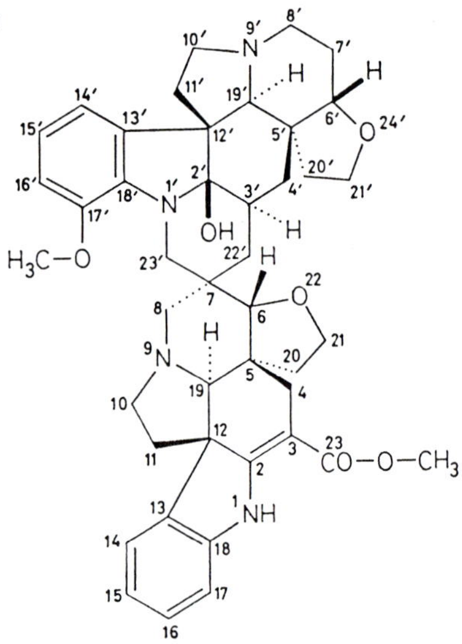

Weinsäure = Tartaric acid

Xanthene **Xanthen**

sys[CA]: 9*H*-Xanthene

$C_{13}H_{10}O$
Beilstein: 5-17/2.252 · Syst.-Nr. 2370
Reg.No.: 92-83-1
IUPAC: B-2.11(9)

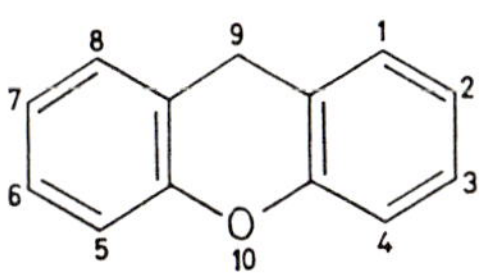

1*H*-Xanthene
CAS-Reg.No.: 27198-07-8
3*H*-Xanthene
CAS-Reg.No.: 261-21-2

Xanthine **Xanthin**

sys[CA]: 1*H*-Purine-2,6-dione, 3,7-dihydro-

$C_5H_4N_4O_2$
Beilstein: 4-26.2327 · Syst.-Nr. 4136
CAS-Reg.No.: 69-89-6

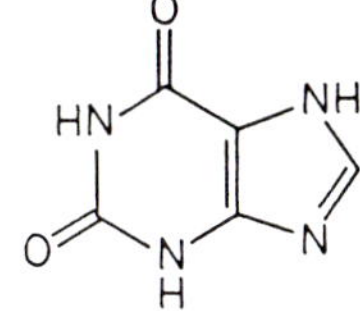

Xanthosine **Xanthosin**

$C_{10}H_{12}N_4O_6$
Beilstein: 4-26.2428 · Syst.-Nr. 4136
CAS-Reg.No.: 146-80-5

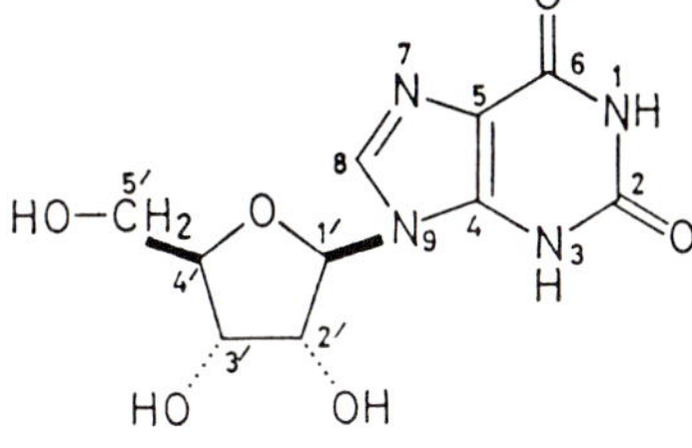

Xylene **Xylol**

sys[CA]: Benzene, dimethyl-

C_8H_{10}
Beilstein · Syst.-Nr. 467
CAS-Reg.No.: 1330-20-7
IUPAC: A-12.1

p-

o-Xylene
Beilstein: 4-5.917
CAS-Reg.No.: 95-47-6
m-Xylene
Beilstein: 4-5.932
CAS-Reg.No.: 108-38-3
p-Xylene
Beilstein: 4-5.951
CAS-Reg.No.: 106-42-3

Xylenol

sys[B + CA]: Phenol, dimethyl-

$C_8H_{10}O$
Beilstein · Syst.-Nr. 529
CAS-Reg.No.: 1300-71-6
IUPAC: C-202.2

Xylenol

[2,3]Xylenol
Beilstein: 4-6.3096
CAS-Reg.No.: 526-75-0

[2,4]Xylenol
Beilstein: 4-6.3126
CAS-Reg.No.: 105-67-9

[2,5]Xylenol
Beilstein: 4-6.3164
CAS-Reg.No.: 95-87-4

[2,6]Xylenol
Beilstein: 4-6.3112
CAS-Reg.No.: 576-26-1

[3,4]Xylenol
Beilstein: 4-6.3099
CAS-Reg.No.: 95-65-8

[3,5]Xylenol
Beilstein: 4-6.3141
CAS-Reg.No.: 108-68-9

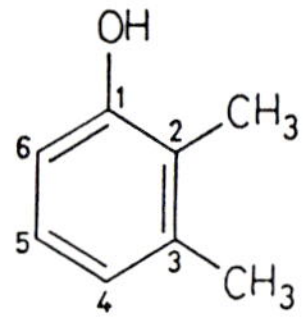

OH
1
2
CH_3
3
CH_3
4
5
6
[2,3]

Xylidine **Xylidin**
sys[B]: Anilin, *ar*-Dimethyl-
[CA]: Benzenamine, *ar*-dimethyl-

$C_8H_{11}N$
Beilstein · Syst.-Nr. 1704
CAS-Reg.No.: 1300-73-8
IUPAC: C-812.1

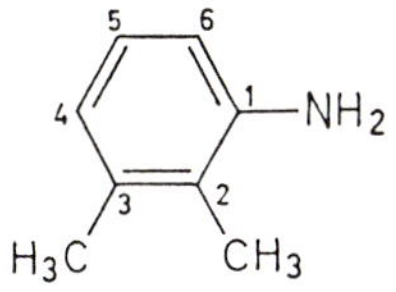

[2,3]

[2,3]Xylidine
Beilstein: 4-12.2497
CAS-Reg.No.: 87-59-2
[2,4]Xylidine
Beilstein: 4-12.2545
CAS-Reg.No.: 95-68-1
[2,5]Xylidine
Beilstein: 4-12.2567
CAS-Reg.No.: 95-78-3
[2,6]Xylidine
Beilstein: 4-12.2521
CAS-Reg.No.: 87-62-7
[3,4]Xylidine
Beilstein: 4-12.2502
CAS-Reg.No.: 95-64-7
[3,5]Xylidine
Beilstein: 4-12.2561
CAS-Reg.No.: 108-69-0

Xylobiose **Xylobiose**

= D-Xylose, O^4-β-D-xylopyranosyl-

Xylofuranose **Xylofuranose**

$C_5H_{10}O_5$
Beilstein · Syst.-Nr. 133
IUPAC: Carb-5/18

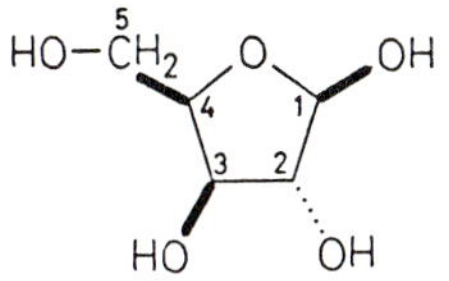

β-D-

α-D-Xylofuranose
CAS-Reg.No.: 14795-83-6
β-D-Xylofuranose
CAS-Reg.No.: 37110-85-3

Xylol = Xylene

Xylopyranose **Xylopyranose**

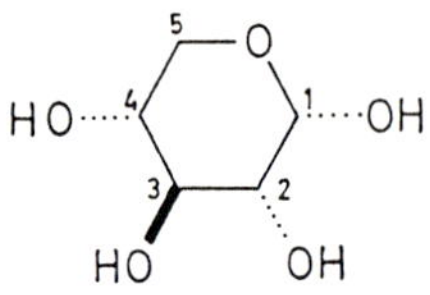

$C_5H_{10}O_5$
Beilstein · Syst.-Nr. 133
IUPAC: Carb-5/18

α-D-Xylopyranose
Beilstein: 4-1.4224
CAS-Reg.No.: 6763-34-4
β-D-Xylopyranose
Beilstein: 4-1.4223
CAS-Reg.No.: 2460-44-8
α-L-Xylopyranose
Beilstein: 4-1.4229
CAS-Reg.No.: 7296-58-4
β-L-Xylopyranose
CAS-Reg.No.: 7322-30-7

α-D-

Xylose **Xylose**

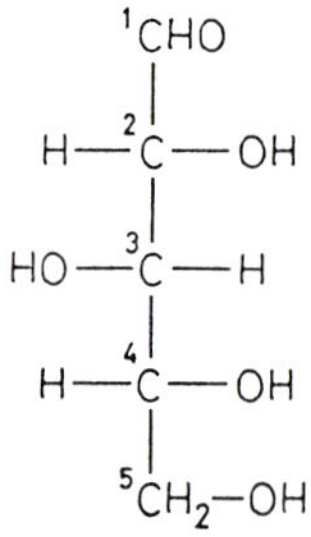

$C_5H_{10}O_5$
Beilstein · Syst.-Nr. 133
CAS-Reg.No.: 25990-60-7
IUPAC: Carb-5

D-Xylose
Beilstein: 4-1.4223
CAS-Reg.No.: 58-86-6
–, O^3-D-xylopyranosyl- = *Rhodymenabiose*
–, O^4-D-xylopyranosyl- = *Xylobiose*

L-Xylose
Beilstein: 4-1.4228
CAS-Reg.No.: 609-06-3
DL-Xylose
Beilstein: 4-1.4229
CAS-Reg.No.: 41247-05-6

D-

Yohimban **Yohimban**

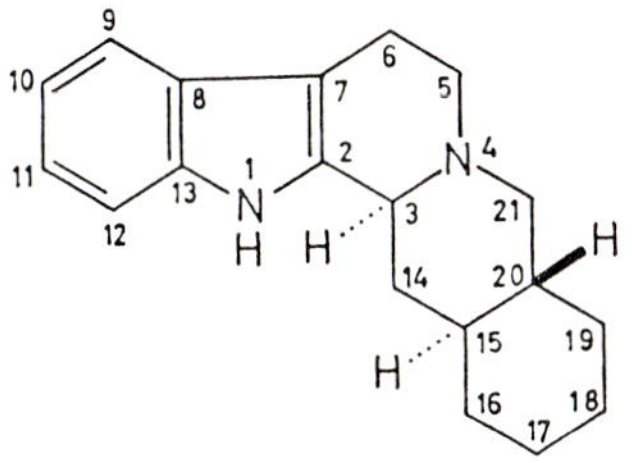

$C_{19}H_{28}N_2$
Beilstein: 4-22.4327* · Syst.-Nr. 3484
CAS-Reg.No.: 523-06-8
IUPAC: F-(44)

–, 17,18-seco- = Corynan

Zimtsäure = Cinnamic acid

Root-Modifying Prefixes

The root-modifying prefixes explained below are predominantly used with steroid and terpene names; they are considered to be part of the root and, unlike other prefixes, they are not seperated from the root when names are inverted for indexing[1].

x (y → z)-Abeo- indicates a rearrangement of the skeleton thus: an x – y bond is broken and a new x – z bond is formed; the original numbering of the skeleton is retained (IUPAC-rule: F-4.9).

"Abeo-" will be used only in exceptional cases in both BEILSTEIN and CA.

Examples:

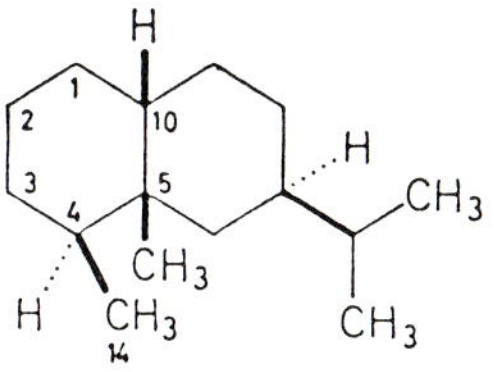

Eremophilane

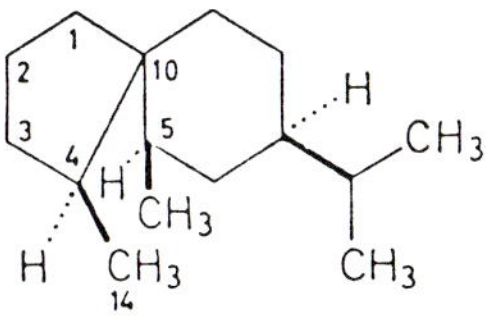

4(5 → 10)-Abeo-eremophilane

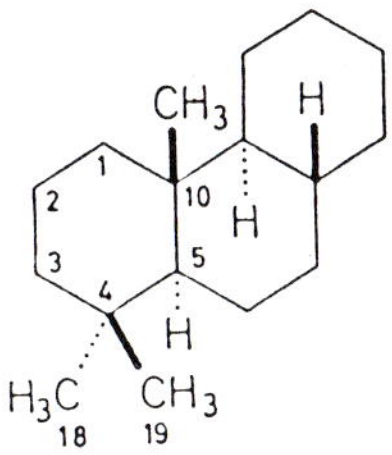

Podocarpane

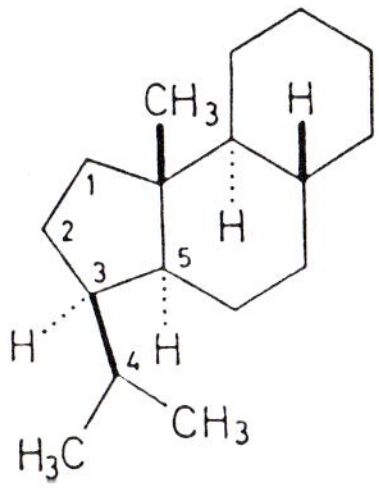

(3α*H*)-5(4 → 3)-Abeo-podocarpane

---

[1] In this book occasional departures have been made from this principle in order not to obscure the relationship with the unchanged roots.

Cyclo- indicates the generation of an extra ring due to the formation of a bond between any two non-adjacent atoms of the basic skeleton. The locants of the two atoms connected by the new bond are placed before the prefix and affixed by a hyphen (IUPAC-rule: F-4.1).

Examples:

Ambrosane 6α,9α-Cyclo-ambrosane

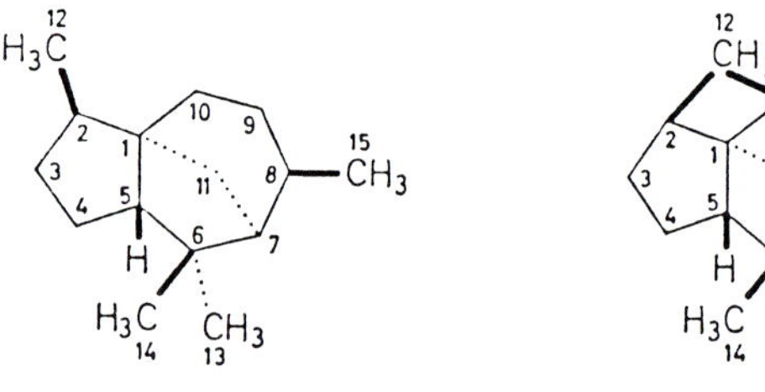

Cedrane 10β,12-Cyclo-cedrane

Friedo- in the pentacyclic terpenes this indicates methyl-group shifts as follows:

D- from 14 to 13
D:C- as in *D*- and in addition from 8 to 14
D:B- as in *D:C*- and in addition from 10 to 9
D:A- as in *D:B*- and in addition from 4 to 5.

The use of "Friedo-" is not recommended by IUPAC (rule: F-4.9, note).

Examples:

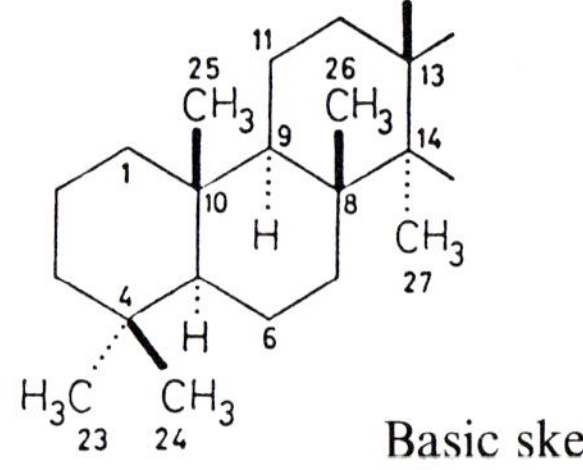

Basic skeleton

D-Friedo-

D:C-Friedo-

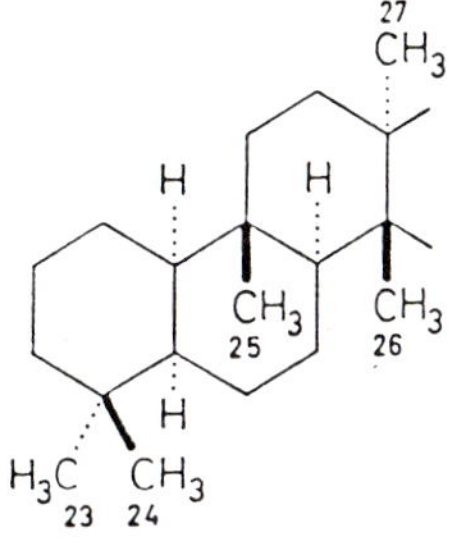

D:B-Friedo-

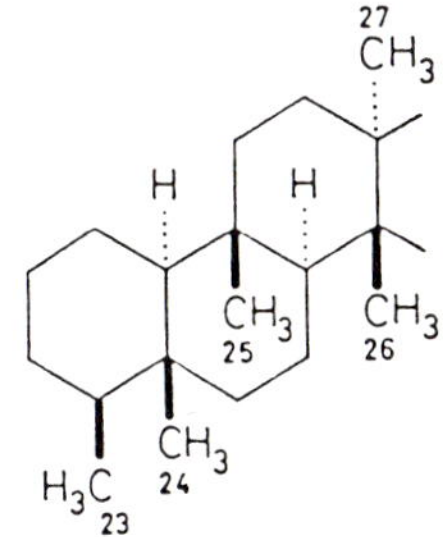

D:A-Friedo-

Homo- indicates an extension of the basic skeleton by means of inserted methylene groups. The method of numbering the inserted atoms is shown in the following examples (IUPAC rule: F-4.5).

Examples:

Taxane

10a-Homo-taxane

2a,10a-Dihomo-taxane

3(8)a-Homo-taxane

Neo- in the pentacyclic terpenes this indicates an alteration in the basic skeleton and the shifting of angular methyl groups as shown in the examples below.

The use of "Neo-" is not recommended by IUPAC (rule: F-4.9, note).

Examples:

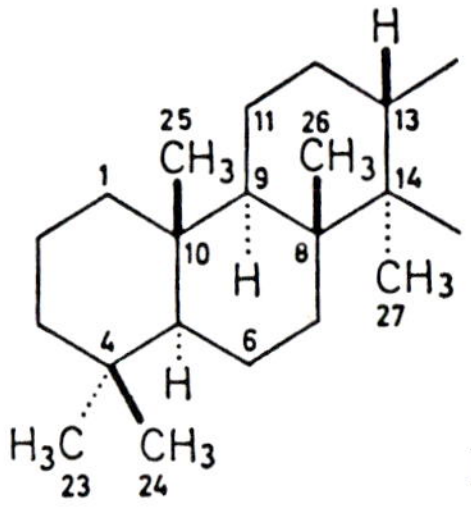

Basic skeleton

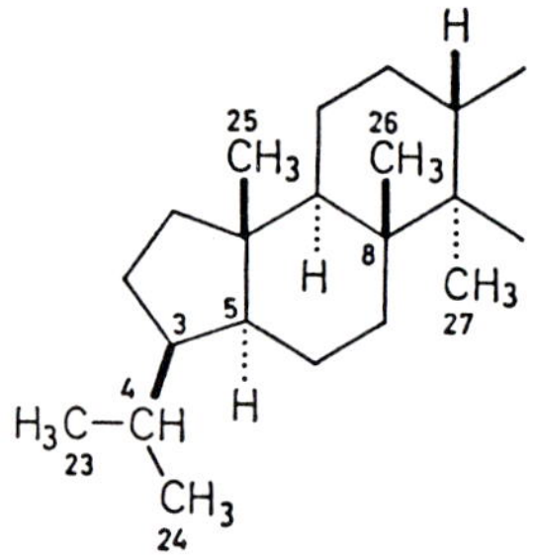

A-Neo-

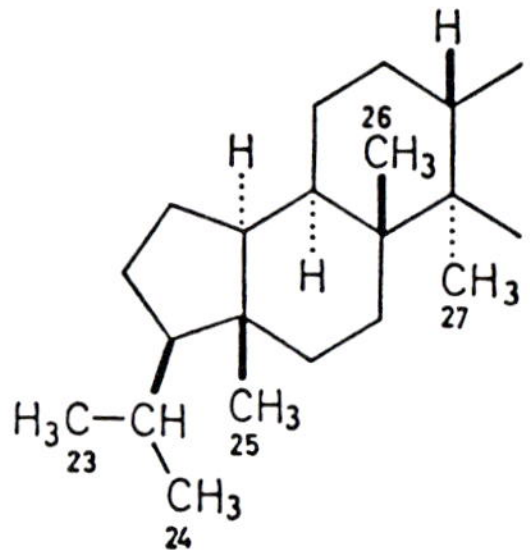

A:B-Neo-

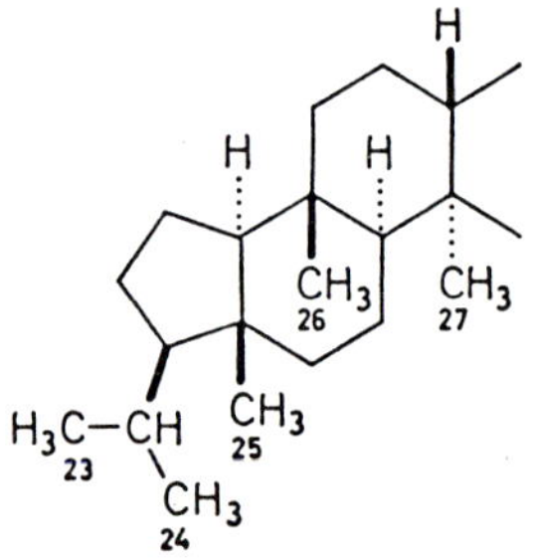

A:C-Neo-

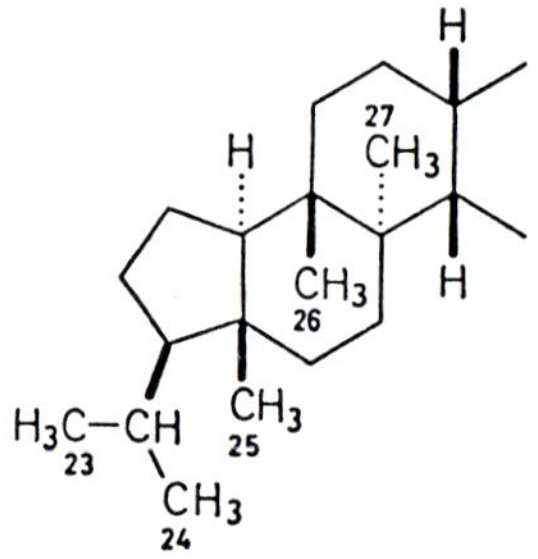

A:D-Neo-

With symmetric triterpenes (e.g. Gammacerane) the following variations are used:

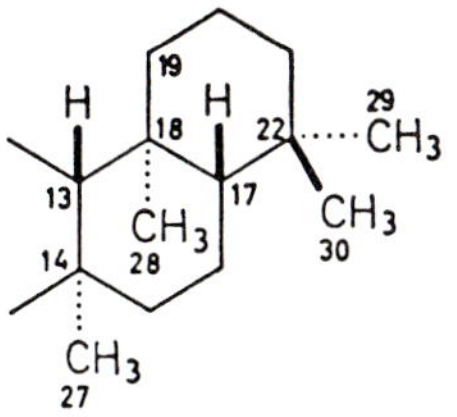

Basic skeleton

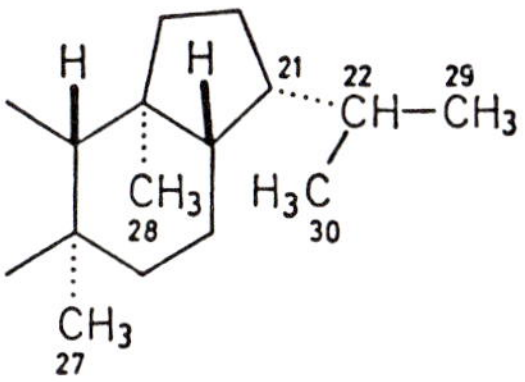

A'-Neo-

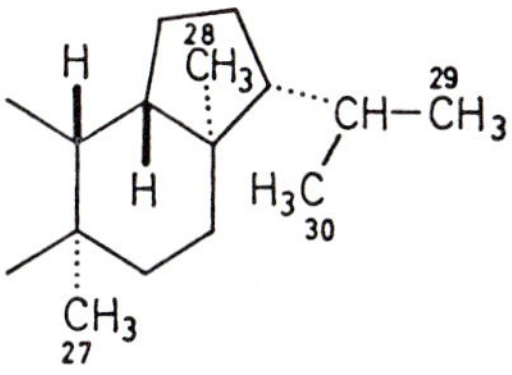

B':*A'*-Neo-

Nor- indicates the removal of methyl- or methylene groups from the basic skeleton. The locant indicates which atom is removed; when alternatives are possible the atom with the highest locant is removed (IUPAC-rule: F-4.2).

Examples:

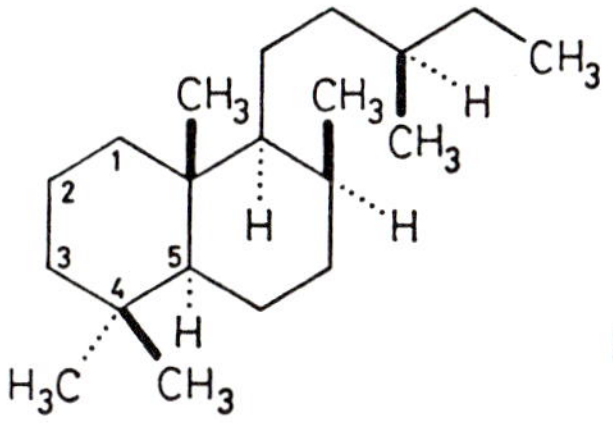

Labdane

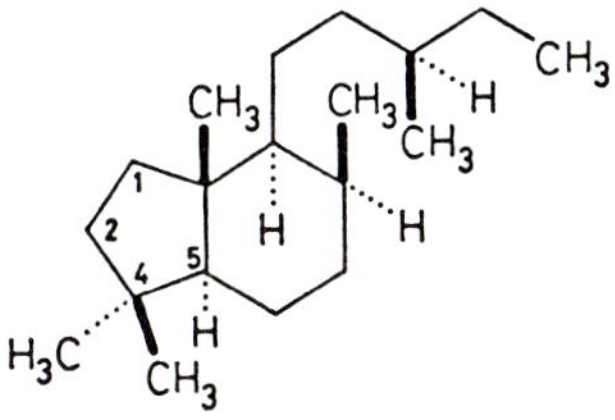

3-Nor-labdane

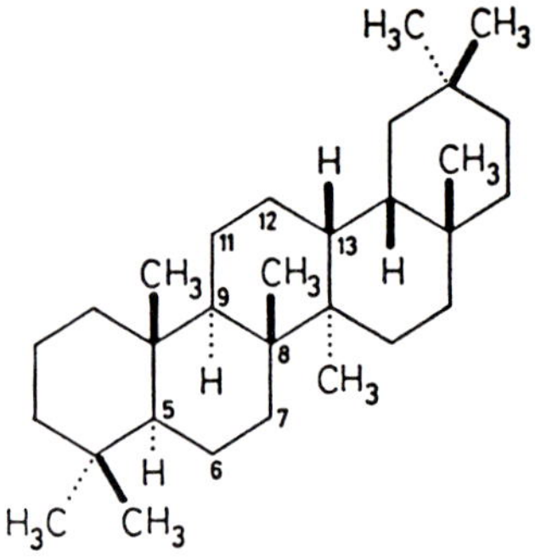

Oleanane 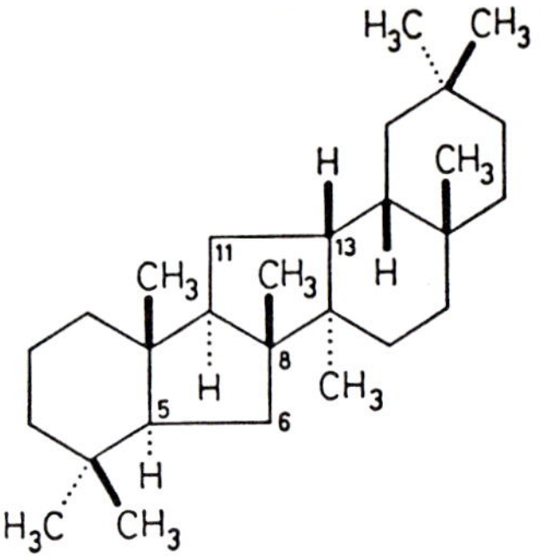 7,12-Dinor-oleanane

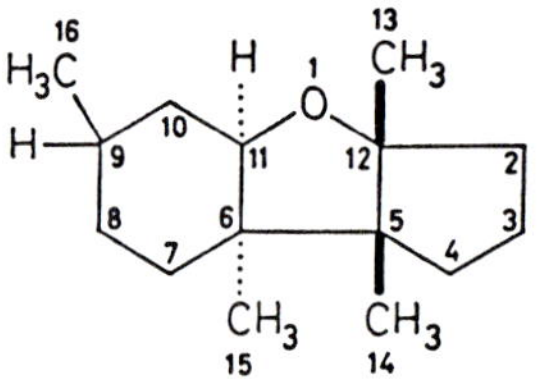

Apotrichothecane 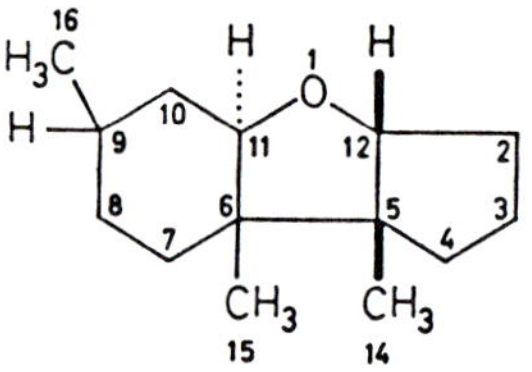 13-Nor-apotrichothecane

Seco- indicates the opening of a ring by the breaking of the bond between the two atoms whose locants are given; the numbering of the basic skeleton is retained (IUPAC-rule: F-4.7).

Examples:

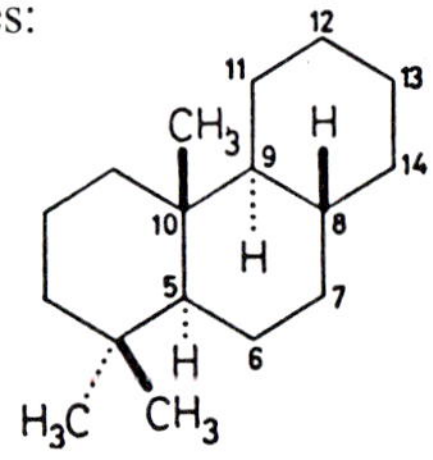

Podocarpane 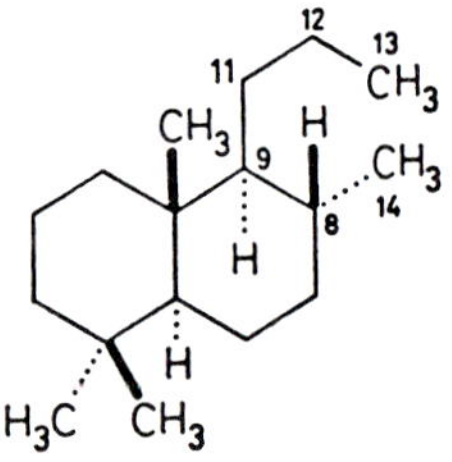 13,14-Seco-podocarpane

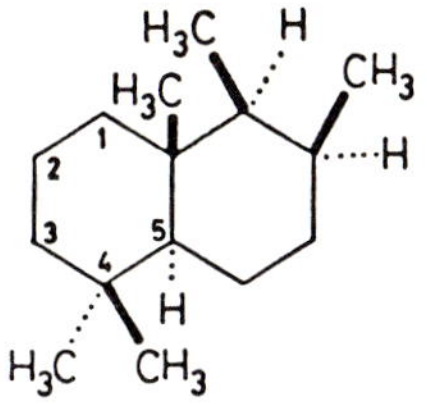

Drimane

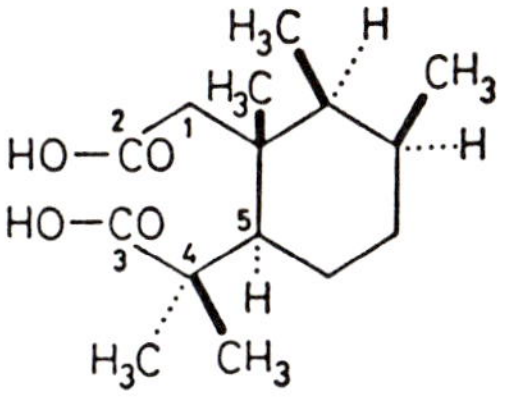

2,3-Seco-driman-2,3-dioic acid

Stammverändernde Präfixe

Die nachfolgend erklärten stammverändernden Präfixe werden überwiegend in der Steroid- und Terpenchemie verwendet. Sie bilden in der Nomenklatur eine Einheit mit dem Stamm [1] und werden bei invertierten Registern nicht vom Stamm abgetrennt.

x (y → z)-Abeo- bezeichnet die Umlagerung des Grundgerüstes in der Weise, daß eine Bindung x – y geöffnet und eine neue Bindung x – z geknüpft wird; die ursprüngliche Bezifferung des Grundgerüstes bleibt erhalten (IUPAC-Regel: F-4.9).

„Abeo-" wird sowohl von BEILSTEIN als auch von CA nur in Ausnahmefällen verwendet.

Beispiele:

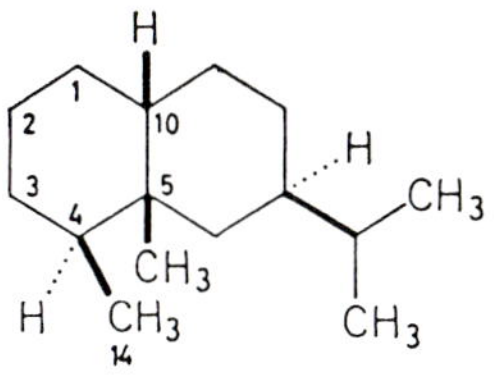

Eremophilan

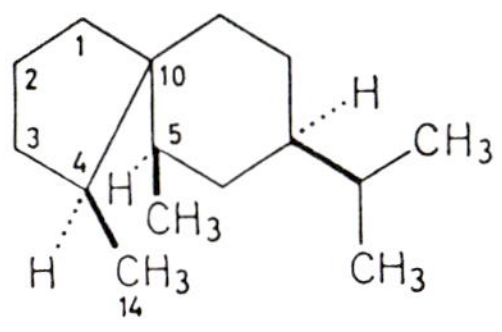

4(5 → 10)-Abeo-eremophilan

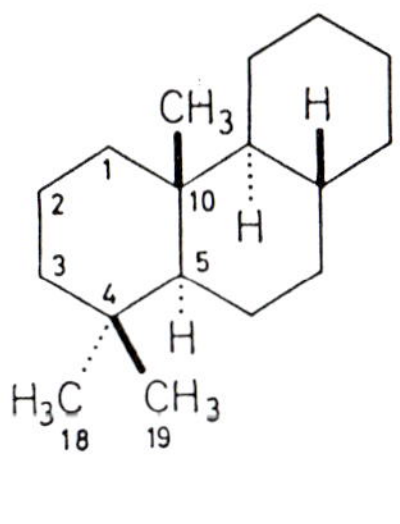

Podocarpan

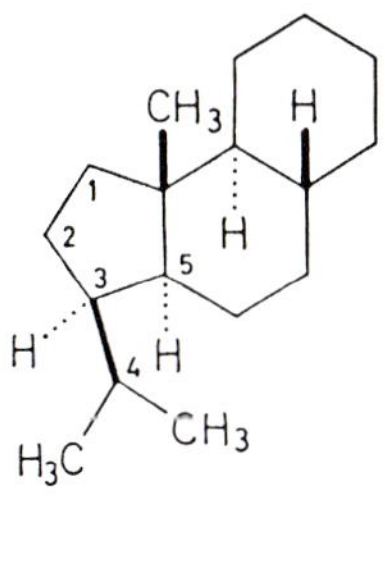

(3α*H*)-5 → 3)-Abeo-podocarpan

[1] Von diesem Prinzip wird im vorliegenden Buch gelegentlich abgewichen, um den Zusammenhang mit den jeweiligen unveränderten Stämmen nicht zu zerstören.

Cyclo- bezeichnet die Bildung eines zusätzlichen Ringes durch Herstellen einer Bindung zwischen zwei beliebigen Atomen des Grundgerüstes. Dem Präfix wird die Bezifferung derjenigen Atome vorangestellt, die durch die neue Bindung verknüpft werden (IUPAC-Regel: F-4.1).

Beispiele:

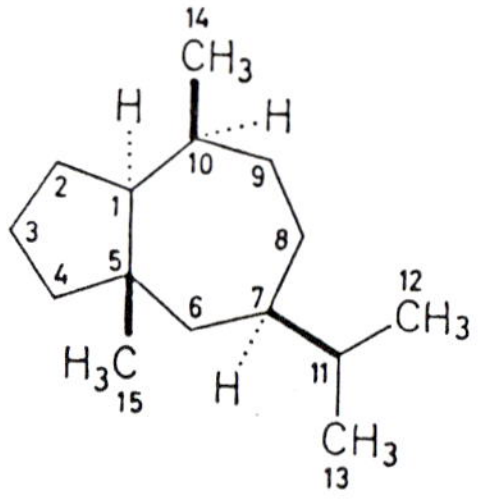

Ambrosan

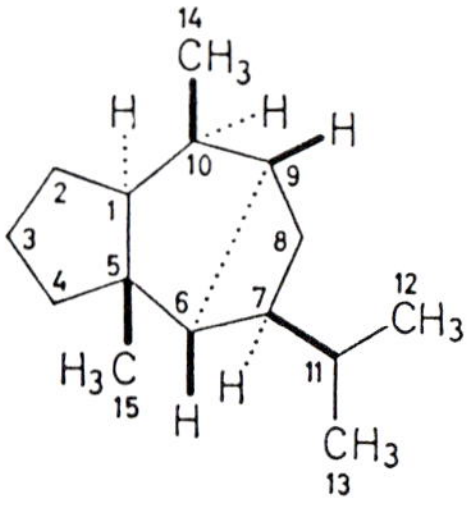

6α,9α-Cyclo-ambrosan

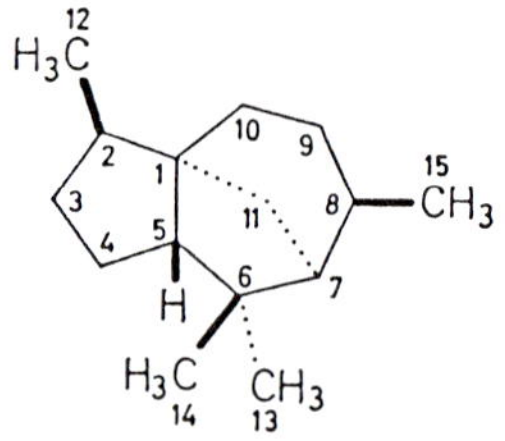

Cedran

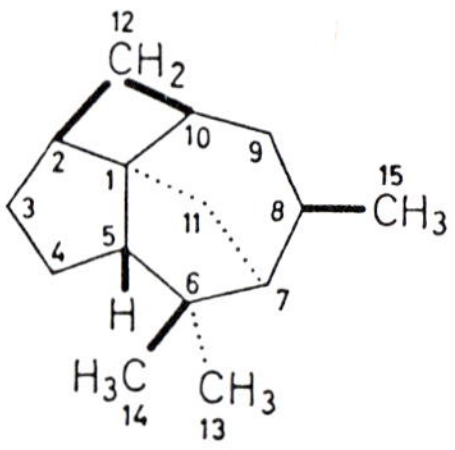

10β,12-Cyclo-cedran

Friedo- bezeichnet bei pentacyclischen Terpenen eine Verschiebung von angulären Methyl-Gruppen, und zwar

D- von 14 nach 13
D:C- wie *D-* und außerdem von 8 nach 14
D:B- wie *D:C-* und außerdem von 10 nach 9
D:A- wie *D:B-* und außerdem von 4 nach 5.

Die Verwendung von „Friedo-“ wird von IUPAC nicht empfohlen (Regel: F-4.9 Anm.).

Beispiele:

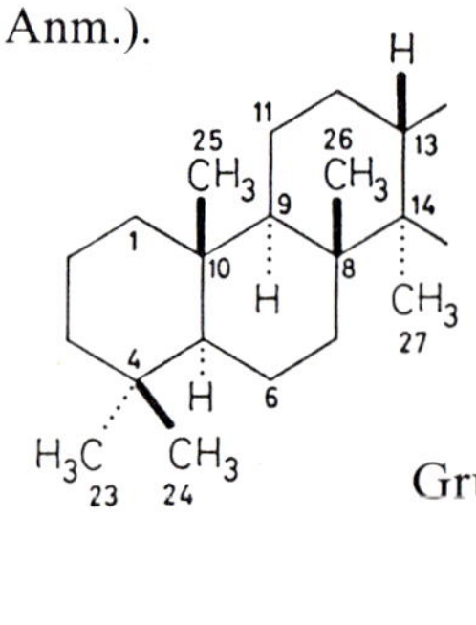

Grundgerüst

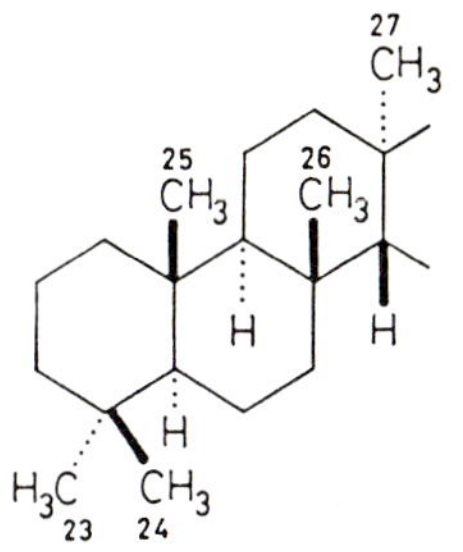

D-Friedo- 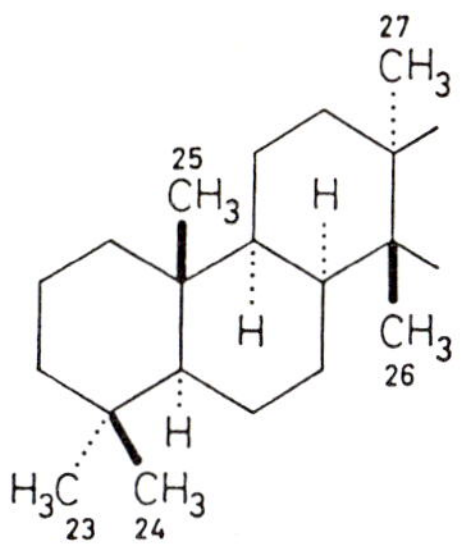 *D:C*-Friedo-

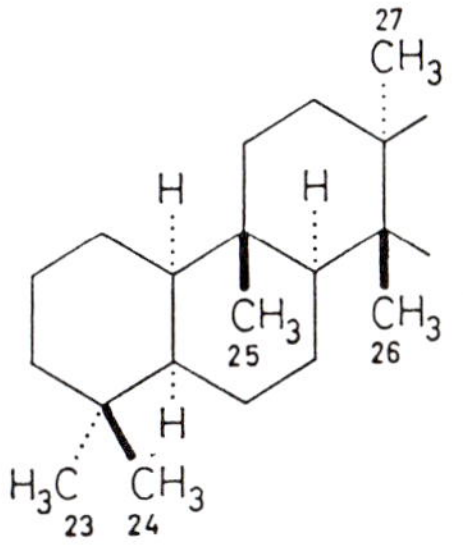

D:B-Friedo- 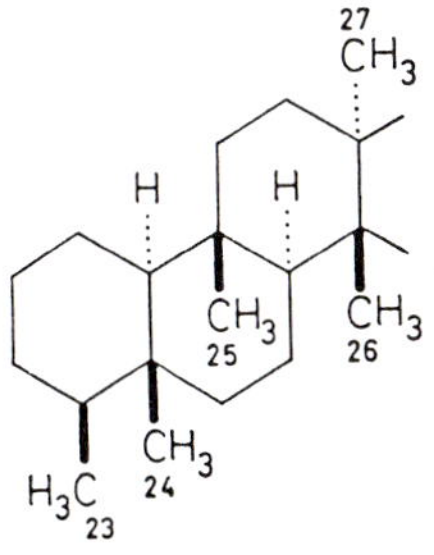 *D:A*-Friedo-

Homo- bezeichnet die Erweiterung des Grundgerüstes durch Einfügen von Methylen-Gruppen. Die Bezifferung der eingefügten Atome ergibt sich aus den folgenden Beispielen (IUPAC-Regel: F-4.5).

Beispiele:

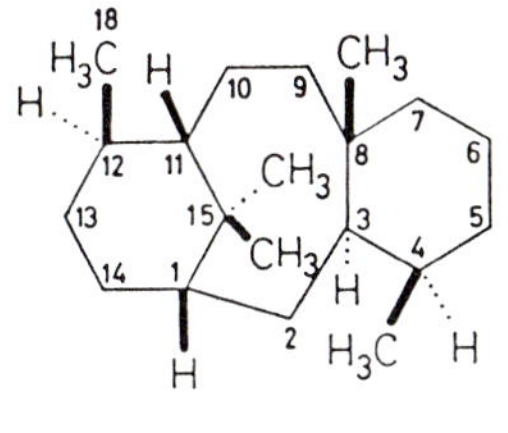

Taxan

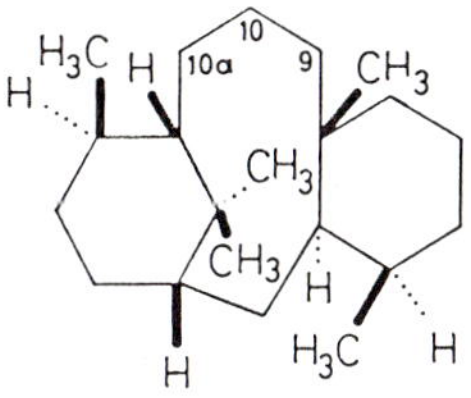

10a-Homo-taxan 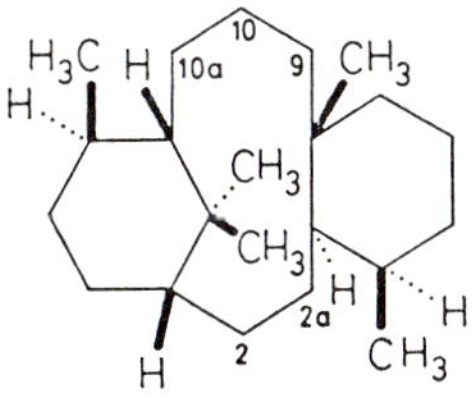 2a,10a-Dihomo-taxan 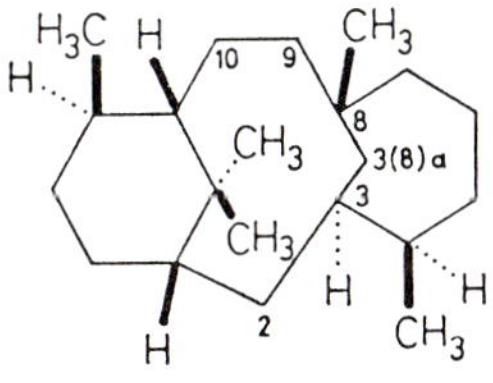 3(8)a-Homo-taxan

Neo- bezeichnet bei pentacyclischen Triterpenen Gerüstveränderungen und eine Verschiebung von angulären Methyl-Gruppen entsprechend den folgenden Beispielen.

Die Verwendung von „Neo-" wird von IUPAC nicht empfohlen (Regel: F-4.9 Anm.).

Beispiele:

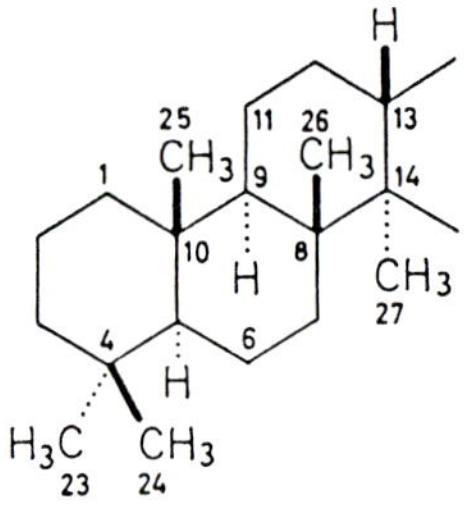

Grundgerüst

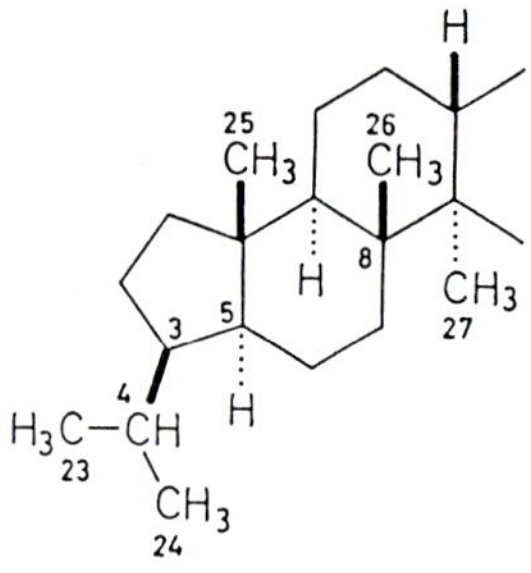

A-Neo-

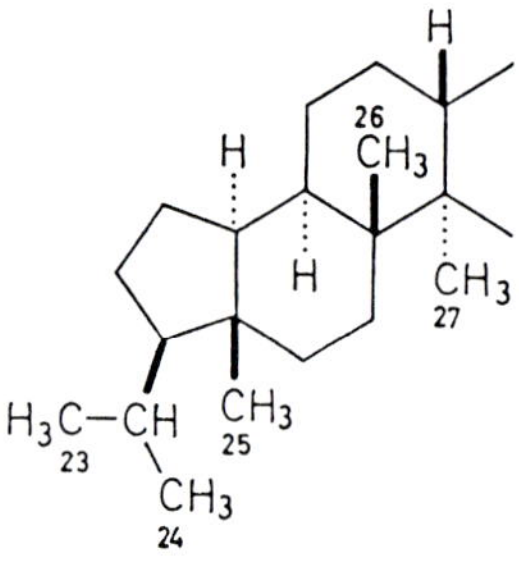

A:*B*-Neo-

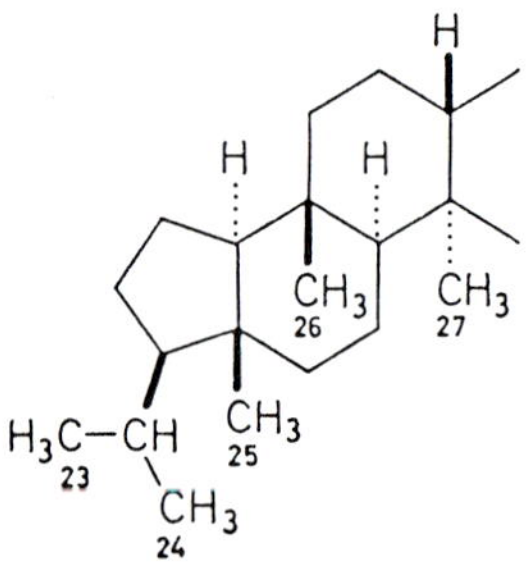

A:*C*-Neo-

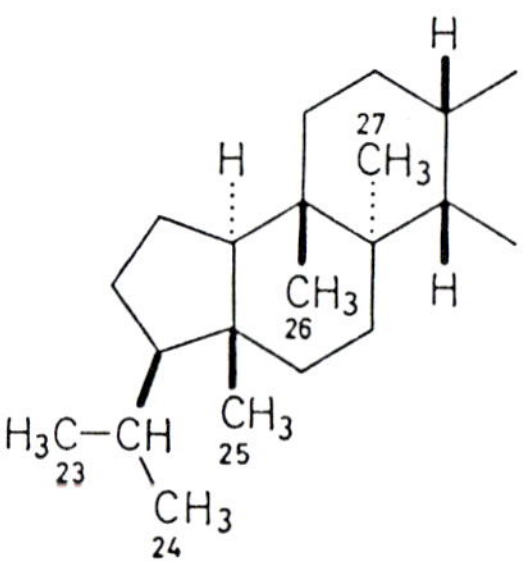

A:*D*-Neo-

Bei symmetrischen Triterpenen (z. B. Gammaceran) werden noch folgende Varianten verwendet:

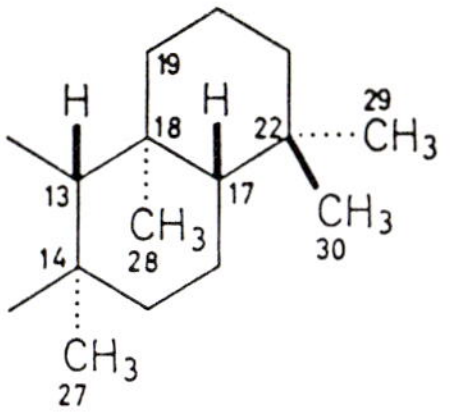

Grundgerüst

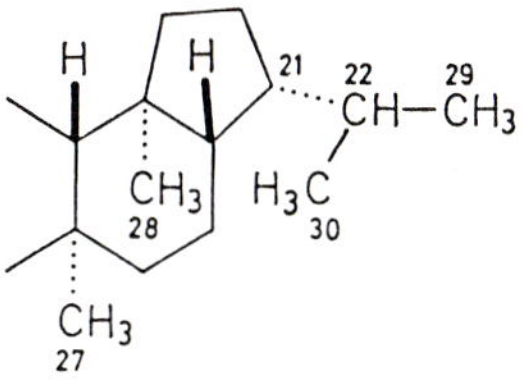

A'-Neo-

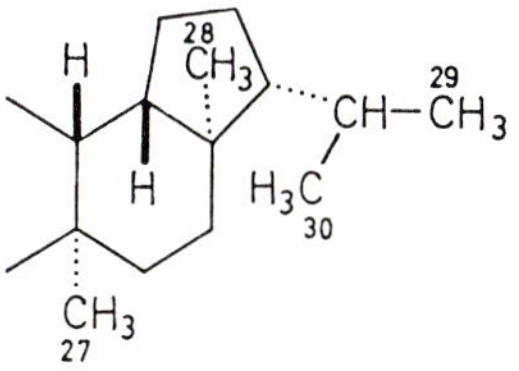

B':*A'*-Neo-

Nor- bezeichnet die Eliminierung von Methyl- oder Methylen-Gruppen aus dem Grundgerüst. Die Bezifferung gibt diejenigen Atome an, die eliminiert werden; sind Alternativen möglich, so entfällt das jeweils höherbezifferte Atom (IUPAC-Regel: F-4.2).

Beispiele:

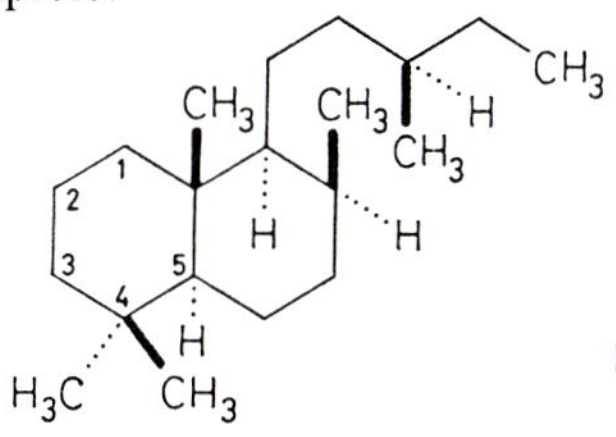

Labdan

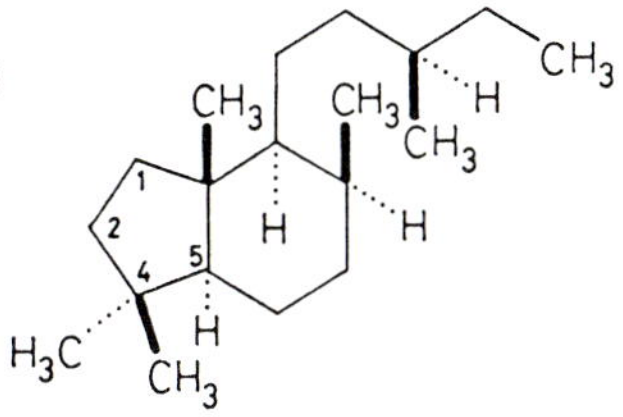

3-Nor-labdan

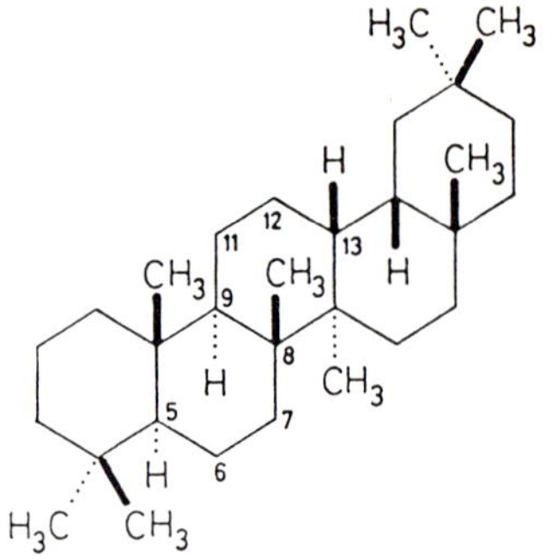

Oleanan

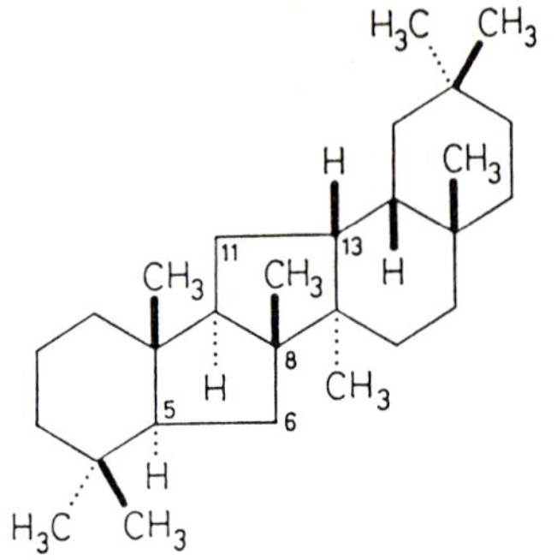

7,12-Dinor-oleanan

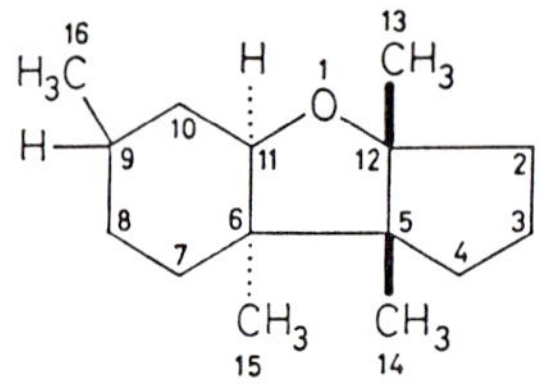

Apotrichothecan

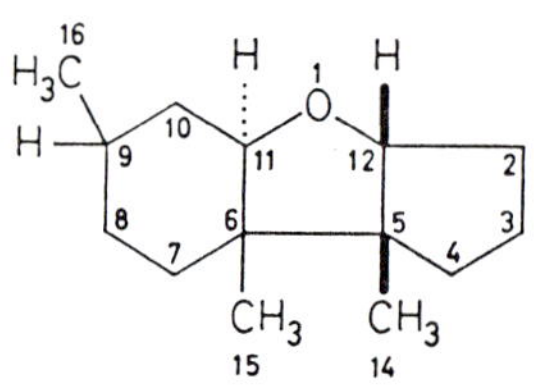

13-Nor-apotrichothecan

Seco- bezeichnet das Öffnen eines Ringes zwischen den durch die Bezifferung angegebenen Atomen; die Bezifferung des Grundgerüstes bleibt erhalten (IUPAC-Regel: F-4.7).

Beispiele:

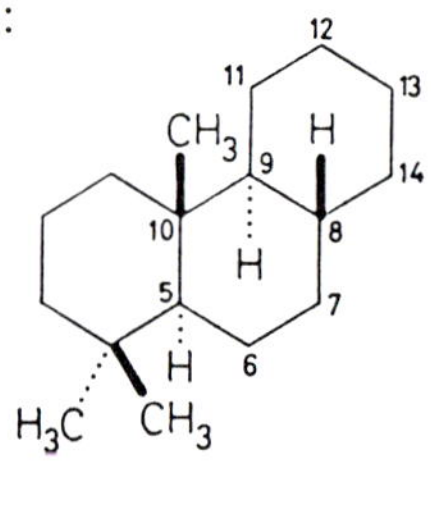

Podocarpan

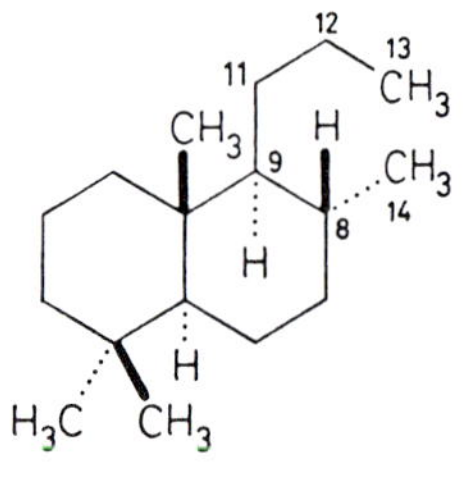

13,14-Seco-podocarpan

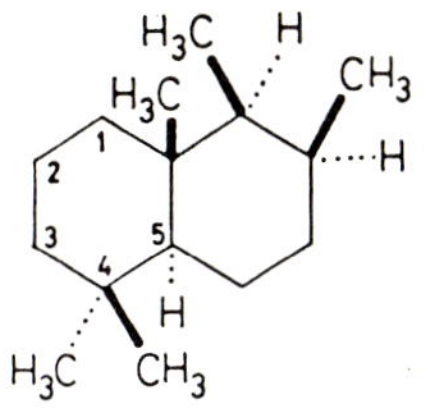

Driman

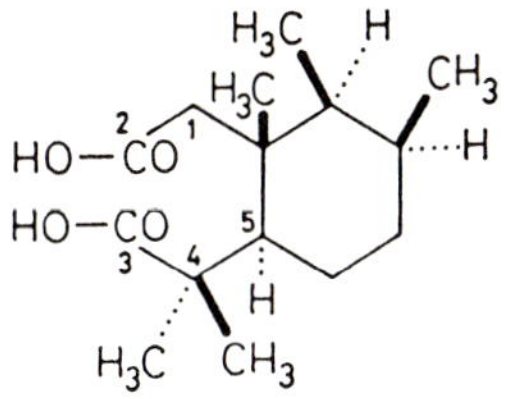

2,3-Seco-driman-2,3-disäure

List of IUPAC-Rules

IUPAC-Commission on the Nomenclature of Organic Chemistry:
Nomenclature of Organic Chemistry

A Section A: Hydrocarbons, 1969

B Section B: Fundamental Heterocyclic Systems, 1969

C Section C: Characteristic Groups Containing Carbon, Hydrogen, Oxygen, Nitrogen, Halogen, Sulfur, Selenium and/or Tellurium, 1969

IUPAC Commission on the Nomenclature of Organic Chemistry and IUPAC-IUB Commission on Biochemical Nomenclature:

Car **Tentative Rules for the Nomenclature of Carotenoids** (1972)

IUPAC Commission on the Nomenclature of Organic Chemistry and IUPAC-IUB Commission on Biochemical Nomenclature:

Carb **Tentative Rules for Carbohydrate Nomenclature** (1969)

IUPAC-Commission on the Nomenclature of Organic Chemistry:
Nomenclature of Organic Chemistry

F Section F: Natural Products and Related Compounds, 1976

IUPAC-IUB Commission on Biochemical Nomenclature:

Fol **Nomenclature and Symbols for Folic acid and Related Compounds** (Tentative Rules, 1965)

IUPAC Commission on the Nomenclature of Organic Chemistry and IUPAC-IUB Commission on Biochemical Nomenclature:

I **Nomenclature of Cyclitols** (Recommendations 1973)

IUPAC-IUB Commission on Biochemical Nomenclature:

Lip **The Nomenclature of Lipids** (Recommendations 1976)

IUPAC-IUB Commission on Biochemical Nomenclature:

M **Trivial Names of Miscellaneous Compounds of Importance in Biochemistry** (1966)

N	IUPAC-IUB Commission on Biochemical Nomenclature: **Abbreviations and Symbols for Nucleic Acids, Polynucleotides and their Constituents** (Recommendations 1970)
Ret	IUPAC-IUB Joint Commission on Biochemical Nomenclature: **Nomenclature of Retinoids** (Recommendations 1981)
Toc	IUPAC-IUB Joint Commission on Biochemical Nomenclature: **Nomenclature of Tocopherols and Related Compounds** (Recommendations 1981)
TP	IUPAC-IUB Joint Commission on Biochemical Nomenclature: **Nomenclature of Tetrapyrroles** (Recommendations 1978)
VB_6	IUPAC-IUB Commission on Biochemical Nomenclature: **Nomenclature for Vitamins B-6 and Related Compounds** (Recommendations 1973)
VD	IUPAC-IUB Joint Commission on Biochemical Nomenclature: **Nomenclature of Vitamin D** (Recommendations 1981)
2S	IUPAC Commission on the Nomenclature of Organic Chemistry and IUPAC-IUB Commission on Biochemical Nomenclature: **The Nomenclature of Steroids** (Revised Tentative Rules 1967)
3AA	IUPAC-IUB Joint Commission on Biochemical Nomenclature: **Nomenclature and Symbolism for Amino Acids and Peptides** (Recommendations 1983)

RingCode-Index/Register

The RingCode-Index is arranged in the same way as that in the "Index of Ring Systems (IRS)" [1]) of Chemical Abstracts that is according to

1. the number of rings in the system,
2. the sizes of the rings,
3. the types and number of ring atoms.

Further subdivision is made according to the number of carbon atoms in the molecule.

Acyclic compounds (arranged in order of the number of carbon atoms) and monocyclic compounds are also included in this list.

Non-condensed ring systems are differentiated from condensed systems by the use of a diagonal stroke instead of commas e.g. "6/6" indicates a system with two isolated 6-rings and "6,6" one with two condensed 6-rings.

Non-condensed ring systems are given precedence over condensed systems containing the same number of rings.

Das folgende RingCode-Register ist entsprechend dem "Index of Ring Systems (IRS)" [1]) von Chemical Abstracts geordnet, d.h.

1) nach der Anzahl der Ringe im System,
2) nach der Grösse der Ringe,
3) nach der Art und Anzahl der Ring-Atome.

Weiter wird nach der Anzahl der C-Atome des Gesamtmoleküls untergliedert.

In dieses Register sind auch die Verbindungen ohne Ring (d.h. die acyclischen Verbindungen, nach ihrer Summenformel geordnet) sowie die Verbindungen mit nur einem Ring aufgenommen worden.

Nicht-kondensierte Ring-Systeme werden von kondensierten Systemen durch einen Schrägstrich statt eines Kommas angezeigt, z.B. "6/6" bezeichnet ein System mit zwei isolierten 6-Ringen, "6,6" ein solches mit zwei kondensierten 6-Ringen. Nichtkondensierte Ring-Systeme werden vor kondensierten Systemen gleicher Ring-Anzahl eingeordnet.

---

[1]) C.A.Index Guide 1984 Appendix II, Section 12

Systems with 0 rings (Acyclic compounds)

C_1 Carbamic acid
Carbazic acid
Carbodiimide
Carbonic acid
Formaldehyde
Formic acid
Guanidine
Isourea
Methane
Semicarbazide
Urea

C_2 Acetaldehyde
Acetic acid
Acetylene
Allophanic acid
Biuret
Dithiobiuret
Ethane
Ethylene glycol
Glycine
Glycolic acid
Glyoxal
Glyoxylic acid
Ketene
Oxalic acid
Oxamic acid
Taurine
Thiobiuret
Thiuram disulfide
Thiuram monosulfide

C_3 Acetone
Acrylic acid
Alanine
β-Alanine
Allene
Cysteine
Glycerol
Glyceric acid
Hydantoic acid
Lactic acid
Malonic acid
Mesoxalic acid
Propane
Propiolic acid
Propionic acid
Pyruvic acid
Sarcosine
Serine
Tartronic acid
Triuret

C_4 Acetoacetic acid
Allothreonine
Asparagine
Aspartic acid
Butane
Butyraldehyde
Butyric acid
Crotonic acid
Erythrose
Fumaric acid
Homocysteine
Homoserine
Isobutane
Maleic acid
Malic acid
Methacrylic acid
Oxalacetic acid
Succinaldehyde
Succinic acid
Tartaric acid
Threonine
Threose

C_5 Apiose
Arabinose
Betaine
Choline
Glutamic acid
Glutamine
Glutaric acid
Isopentane
Isoprene
Isovaline
Lyxose
Methionine
Neopentane

Norvaline
Ornithine
Pentaerythritol
Ribose
Valeric acid
Valine
Xylose

C_6 Adipic acid
Alloisoleucine
Allose
Altrose
Arginine
Cystine
Fructose
Galactose
Glucose
Gulose
Idose
Isohexane
Isoleucine
Lanthionine
Leucine
Lysine
Mannose
Norleucine
Psicose
Sorbose
Tagatose
Talose

C_7 Cystathionine
Sedoheptulose

C_9 Muramic acid
Neuraminic acid

C_{10} Citral
Geraniol

C_{18} Elaidic acid
Oleic acid
Sphinganine
Sphingosine

C_{20} Phytol

C_{30} Squalene

C_{40} ψ,ψ-Carotene

Systems with 1 ring

5

C_2N_2O
C_2 Furazan

C_3NO
C_3 Isoxazole
Isoxazolidine
Rhodanine

C_3NS
C_3 Isothiazole

C_3N_2
C_3 Hydantoin
Imidazole
Imidazolidine
Imidazoline
Pyrazole
Pyrazolidine
Pyrazoline
C_6 Histidine

C_4N
C_4 Pyrrole
Pyrrolidine
Pyrroline
C_5 Proline

C_4O
C_4 Furan
C_5 Arabinofuranose
Lyxofuranose
Ribofuranose
Xylofuranose
C_6 Allofuranose
Altrofuranose
Ascorbic acid
Fructofuranose
Galactofuranose
Glucofuranose

Gulofuranose
Idofuranose
Mannofuranose
Psicofuranose
Sorbofuranose
Tagatofuranose
Talofuranose

C_4S
C_4 Thiophene

C_5
C_6 Fulvene
C_{10} Camphoric acid
Iridane
C_{20} Prostane

6

C_4NO
C_4 Morpholine

C_4N_2
C_4 Alloxan
Barbituric acid
Cytosine
Piperazine
Pyrazine
Pyridazine
Pyrimidine
Uracil
C_5 Orotic acid
Thymine

C_5N
C_5 Piperidine
Pyridine
C_6 Isonicotinic acid
Nicotinic acid
C_8 Pyridoxal
Pyridoxamine
Pyridoxine

C_5O
C_5 Arabinopyranose
Lyxopyranose
Pyran
Ribopyranose
Xylopyranose
C_6 Allopyranose
Altropyranose
Fructopyranose
Galactopyranose
Glucopyranose
Gulopyranose
Idopyranose
Mannopyranose
Psicopyranose
Sorbopyranose
Tagatopyranose
Talopyranose
C_{20} Thromboxane

C_6
C_6 Aniline
Benzene
Hydroquinone
Inositol
Phenol
Phloroglucinol
Picric acid
Pyrocatechol
Pyrogallol
Quercitol
Resorcinol
Streptamine
Styphnic acid
Sulfanilic acid
C_7 Anisidine
Anisole
Anthranilic acid
Benzoic acid
Benzyl alcohol
Cresol
Gallic acid
Guaiacol
Protocatechuic acid
Quinic acid
Salicylic acid
Toluene
C_8 Anisic acid
Isophthalic acid
Phenethyl alcohol
Phenetidine

Phenetole
Phthalic acid
Styrene
Terephthalic acid
Vanillic acid
Vanillin
Veratrole
Xylene
Xylenol
Xylidine
C_9 Cinnamic acid
Cumene
Hippuric acid
Mesitylene
Phenylalanine
Tropic acid
Tyrosine
Veratric acid
C_{10} Anethole
Carvacrol
Cymene
Eugenol
Menthane
Menthol
Thymol
C_{20} Deoxyretinol
Lobane
Retinal
Retinol
Retinoic acid
C_{40} β,ψ-Carotene
ε,ψ-Carotene
φ,ψ-Carotene

7

C_6N
C_6 ε-Caprolactam

10

C_{10}
C_{15} Germacrane

16

$C_{15}O$
C_{20} Leuconolide

Systems with 2 rings

3,5

C_3-C_5
C_{10} Thujane

3,6

C_3-C_6
C_7 Norcarane
C_{10} Carane

3,14

C_2O-$C_{13}O$
C_{20} Oleandolide

4,5

C_3N-C_3NS
C_8 Penicillanic acid

4,6

C_4-C_6
C_7 Norpinane
C_{10} Pinane

5/5

C_5/C_5
C_{40} κ,κ-Carotene

5,5

C_5-C_5
C_7 Norbornane
C_8 Pentalene
C_{10} Bornane
Camphene

5/6

C_4N/C_5N
C_{10} Nicotine

C_4O/C_4N_2
C_9 Uridine
C_{10} Orotidine
Thymidine

C_5/C_6
C_{40} β,κ-Carotene
κ,φ-Carotene
κ,χ-Carotene

5,6

C_3N_2-C_4N_2
C_5 Adenine
Guanine
Hypoxanthine
Purine
Xanthine

C_3N_2-C_6
C_7 Indazole

C_3O_2-C_6
C_8 Piperonal
Piperonylic acid
C_{10} Isosafrole
Safrole
C_{12} Piperonic acid

C_4N-C_5N
C_8 Indolizine
Tropane

C_4N-C_6
C_8 Indole
Indoline
Isoindole
Isoindoline
C_{11} Tryptophane

C_4O-C_4N_2
C_9 Cytidine

C_4O-C_6
C_8 Isobenzofuran
Phthalan
Phthalide

C_5-C_6
C_9 Indan
Indene
C_{15} Bakkane

5,7

C_5-C_7
C_{10} Azulene
C_{15} Ambrosane
Guaiane

6/6

C_5N/C_5N
C_{10} Anabasine

C_6/C_6
C_{12} Benzidine
Biphenyl
C_{14} Benzil
Benzilic acid
Deoxybenzoin
Stilbene
C_{15} Chalcone
Thyronine
C_{40} β,β-Carotene
β,ε-Carotene
β,φ-Carotene
β,χ-Carotene
ε,ε-Carotene
φ,φ-Carotene
φ,χ-Carotene
χ,χ-Carotene

6,6

C_4N_2-C_4N_2
C_6 Pteridine

C_4N_2-C_6
C_8 Cinnoline
Phthalazine
Quinazoline
Quinoxaline

C_5N-C_5N
C_7 Quinuclidine
C_8 Naphthyridine
C_9 Quinolizine

C_5N-C_6
C_9 Isoquinoline
Quinoline

C_5O-C_6
C_9 Chroman
Chromene
Coumarin
Isochroman
Isochromene
Isocoumarin
C_{26} Tocol

C_6-C_6
C_{10} Naphthalene
Naphthionic acid
Naphthol
C_{11} Naphthoic acid
C_{15} Cadinane
Drimane
Eremophilane
Eudesmane
C_{20} Clerodane
Labdane

7,7

C_7-C_7
C_{12} Heptalene

Systems with 3 rings

3,3,3

C_3-C_3-C_3
C_4 Tetrahedrane

3,7,7

C_3-C_7-C_7
C_{10} Bullvalene

4,6,6

C_4-C_6-C_6
C_{12} Biphenylene

4,6,7

C_4-C_6-C_7
C_{15} Caryolane

5/5/5

$C_4N/C_4N/C_4N$
C_{14} Tripyrrin

5/5,6

C_4O/C_3N_2-C_4N_2
C_{10} Adenosine
Guanosine
Inosine
Xanthosine

5,5,6

C_4O-C_5-C_6
C_{15} Apotrichothecane

C_5-C_5-C_6
C_{12} *as*-Indacene
s-Indacene
C_{15} Cedrane

5,5,7

C_5-C_5-C_7
C_{15} Longifolane

5,5,8

C_5-C_5-C_8
C_{25} Ophiobolane

5,5,11

C_4N-C_4N-C_9O_2
C_{17} Crotalanan

5,5,12

C_4N-C_4N-$C_{10}O_2$
C_{18} Senecionan

5,5,14

C_4N-C_4N-$C_{11}O_3$
C_{22} Parsonsine

5,6/6

C_5-C_6/C_6
C_{27} Calciol
Isocalciol
Tacalciol
C_{28} Ercalciol

5,6,6

C_4N-C_5N-C_6
C_{11} β-Carboline

C_4N-C_6-C_6
C_{12} Carbazole

C_5-C_5O-C_6
C_{15} Trichothecane

C_5-C_6-C_6
C_{12} Acenaphthene
Acenaphthylene
C_{13} Fluorene
C_{15} Clovane

5,6,7

C_5-C_6-C_7
C_{20} Clavularane

5,6,11

C_4N-C_6-C_{11}
C_{20} [11]Cytochalasan

5,6,13

C_4N-C_6-C_{13}
C_{22} [13]Cytochalasan

6/6,6

C_6/C_4N_2-C_4N_2
C_{14} Pteroic acid

C_6/C_6-C_6
C_{30} Ambrane

6,6,6

C_4As_2-C_6-C_6
C_{12} Arsanthrene

C_4AsN-C_6-C_6
C_{12} Phenarsazine

C_4NO-C_6-C_6
C_{12} Phenoxazine

C_4NS-C_6-C_6
C_{12} Phenothiazine

C_4N_2-C_4N_2-C_6
C_{17} Riboflavin

C_4N_2-C_6-C_6
C_{11} Perimidine
C_{12} Phenazine

C_4OS-C_6-C_6
C_{12} Phenoxathiin

C_4S_2-C_6-C_6
C_{12} Thianthrene

C_5As-C_6-C_6
C_{13} Acridarsine
Arsanthridine

C_5N-C_5N-C_5N
C_{11} Anthyridine
Cytisine

C_5N-C_5N-C_6
C_{12} Phenanthroline

C_5N-C_6-C_6
C_{13} Acridine
Phenanthridine

C_5O-C_6-C_6
C_{13} Xanthene

C_6-C_6-C_6
C_{10} Adamantane
C_{13} Phenalene
C_{14} Anthracene
Phenanthrene
C_{17} Podocarpane
C_{20} Abietane
Pimarane
Rosane

6,6,8

C_6-C_6-C_8
C_{20} Taxane

6,6,26

C_5N-C_5N-$C_{22}N_2O_2$
C_{28} Carpaine

Systems with 4 rings

3,5,6,7

C_3-C_5-C_6-C_7
C_{20} Tigliane

3,6,6,19

C_2O-C_4NO-C_6-$C_{18}N$
C_{34} Maytansine

5/5/5/5

$C_4N/C_4N/C_4N/C_4N$
C_{19} Bilin
C_{33} Bilirubin
Biliverdin
Mesobilirubin
Mesobiliverdin
Stercobilin
Urobilin
Urobilinogen

5,5,5,6

C_4N-C_4O-C_5-C_6
C_{16} Dendrobane

C_4N-C_5-C_5-C_6
C_{19} Acutumine

C_5-C_5-C_5-C_6
C_{15} Trindene

5,5,6,6

C_4N-C_4N-C_5N-C_6
C_{13} Securinane
C_{19} Corynoxan

C_5-C_5-C_6-C_6
C_{15} Gibbane

5,5,6,7

C_5-C_5-C_6-C_7
C_{20} Grayanotoxane

5/6,6,6

C_4O/C_5O-C_6-C_6
C_{26} Obacunoic acid

5,6,6,6

C_4N-C_5N-C_5N-C_6
C_{19} Corynan

C_4N-C_5N-C_5O-C_6
C_{16} Eleocarpine

C_4N-C_5N-C_6-C_6
C_{16} Erythrinan

C_4N-C_5N-C_6-C_6
C_{14} Ergoline
C_{15} Galanthan
Quindoline
Quinindoline
C_{17} Dasycarpidan

C_4N-C_6-C_6-C_6
C_{16} Hasubanan

C_4O-C_5O-C_6-C_6
C_{15} Lycoranan

C_5-C_6-C_6-C_6
C_{16} Aceanthrene
Aceanthrylene
Acephenanthrene
Acephenanthrylene
Fluoranthene
C_{17} Gonane
C_{18} Estrane
C_{19} Androstane
C_{20} Kaurane
Phyllocladane
C_{21} Diginane
Pregnane
C_{24} Cholane
C_{27} Cholestane
Jervane
C_{28} Ergostane
Lumistane
C_{29} Stigmastane
C_{30} Cucurbitane
Dammarane
Euphane
Lanostane
Tirucallane
C_{31} Eburicane
Euphorbane

5,6,6,7

C_4O-C_6-C_6-C_6N
C_{15} Galanthaman

C_5-C_5N-C_6-C_6N
C_{16} Fawcettidine

5,6,6,8

C_4N-C_5N-C_6-C_8
C_{20} Vobasan

5,6,6,20

C_4O-C_6-C_6-$C_{17}N_3$
C_{25} Lunaridine
Lunarine

6,6/6,6

C_5N-C_5N/C_5N-C_6
  C_{19} Cinchonan

C_6-C_6/C_6-C_6
  C_{30} Onocerane

6,6,6,6

C_5O-C_6-C_6-C_6
  C_{20} Picrasane

C_5N-C_5N-C_5N-C_5N
  C_{15} Matridine
  Sparteine

C_5N-C_5N-C_6-C_6
  C_{15} Lycopodane
  C_{17} Berbine

C_5N-C_6-C_6-C_6
  C_{15} Thebenidine

C_6-C_6-C_6-C_6
  C_{16} Pyrene
  C_{18} Chrysene
  Naphthacene
  Triphenylene
  C_{20} Atisane

6,6,6,7

C_5O-C_6-C_6-C_6N
  C_{17} Rheadan

C_6-C_6-C_6-C_7
  C_{18} Pleiadene

6,6,6,23

C_4O_2-C_6-C_6-$C_{22}N$
  C_{39} Streptovaricinoic acid

Systems with 5 rings

3,5,5,6,6

C_2O-C_4O-C_4O-C_6-C_6
  C_{26} Limonoic acid

3,5,6,6,6

C_3-C_5-C_6-C_6-C_6
  C_{30} Gorgostane

3,6,6/5,5

C_2O-C_6-C_6/C_4O-C_4O
  C_{24} Caryoptinol

4,4,4,4,4

C_4-C_4-C_4-C_4-C_4
  C_8 Cubane

5,5,5,5,15

C_4N-C_4N-C_4N-C_4N-$C_{11}N_4$
  C_{19} Corrin
  Corrole

5,5,5,5,16

C_4N-C_4N-C_4N-C_4N-$C_{12}N_4$
  C_{20} Bacteriochlorin
  Chlorin
  Porphyrin
  Porphyrinogen
  C_{30} Deuteroporphyrin
  C_{31} Pyrrochlorin
  Pyrroporphyrin
  C_{32} Etioporphyrin I
  Phyllochlorin

Phylloporphyrin
Rhodochlorin
Rhodoporphyrin
C_{34} Hematoporphyrin
Mesoporphyrin
Protoporphyrin
C_{36} Coproporphyrin I
C_{40} Uroporphyrin I
C_{49} Cytoporphyrin

5,5,5,5,30

C_4O-C_4O-C_4O-C_4O-$C_{24}O_8$
C_{40} Nonactin

5/5,6,6,6

C_4O/C_5-C_6-C_6-C_6
C_{23} Cardanolide

5,5,6,6,6

C_3O_2-C_4N-C_5N-C_6-C_6
C_{16} Crinan
Pancracine

C_3O_2-C_4N-C_5O-C_6-C_6
C_{18} Tazettine

C_4N-C_4N-C_5N-C_5O-C_6
C_{18} Formosanan

C_4N-C_4N-C_5N-C_6-C_6
C_{18} Condyfolan
C_{19} Aspidospermidine
Curan

C_4N-C_5-C_5N-C_5N-C_6
C_{19} Meloscine

C_4N-C_5-C_6-C_6-C_6
C_{22} Conane

C_4O-C_5-C_6-C_6-C_6
C_{27} Furostan

C_5-C_5-C_6-C_6-C_6
C_{27} Fesane

5,5,6,6,7

C_4N-C_4N-C_5O-C_6-C_6O
C_{19} Gelsedine

5,5,6,6,8

C_4N-C_5-C_5N-C_6-C_7N
C_{22} Daphnane

5/6,6,6,6

C_4O/C_5O-C_6-C_6-C_6
C_{26} Gedunan

5,6,6,6/6

C_5-C_6-C_6-C_6/C_5N
C_{27} Veratraman
Veratran

C_5-C_6-C_6-C_6/C_5O
C_{24} Bufanolide

5,6,6,6,6

C_4N-C_4N_2-C_6-C_6-C_6
C_{18} Phthaloperine

C_4N-C_5N-C_5N-C_5N-C_6
C_{19} Eburnamenine
Sarpagane

C_4N-C_5N-C_5N-C_5O-C_6
C_{21} Alstophyllan
Talbotine

C_4N-C_5N-C_5N-C_6-C_6
C_{19} Akuammilan
Yohimban

C_4N-C_5N-C_6-C_6-C_6
C_{19} Acrindoline

C_5-C_5N-C_6-C_6-C_6
C_{19} Veatchane

C_5-C_6-C_6-C_6-C_6
C_{20} Cholanthrene
C_{30} Lupane

5,6,6,6,7

C_4N-C_5N-C_6-C_6-C_6N
C_{19} Ibogamine

5,6,6,6,14

C_4O-C_5N-C_6-C_6-$C_{11}O_3$
C_{36} Evonine

5,6,6,6,15

C_4O-C_5N-C_6-C_6-$C_{12}O_3$
C_{36} Evonimine

6/6,6,6,6

C_5N/C_5N-C_5N-C_5N-C_6
C_{20} Ormosanine

6,6/6,6,6

C_5N-C_6/C_5N-C_5N-C_6
C_{25} Emetan

6,6,6,6,6

C_5N-C_6-C_6-C_6-C_6
C_{19} Atidane

C_6-C_6-C_6-C_6-C_6
C_{20} Perylene
C_{22} Picene
C_{30} Friedelane
Gammacerane
Oleanane
Taraxastane
Taraxerane
Ursane

6,6,6,6,12

C_5N-C_5N-C_6-C_6-$C_{11}O$
C_{24} Lythran

Systems with 6 rings

3,4,5,6,6,6

C_2O-C_4-C_4O-C_5N-C_5N-C_6
C_{16} Annotinine

5,5,5,5,5,16

C_4N-C_4N-C_4N-C_4N-C_5-$C_{12}N_4$
C_{22} Phorbine
C_{33} Phytochlorin
Phytoporphyrin
C_{35} Pheophorbide a
Pheophorbide b

5,5,5,5,5,19

C_4N-C_4N-C_4N-C_4N-C_4N-$C_{14}N_5$
C_{24} Sapphyrin

5,5,5,6,6,6

C_3O_2-C_3O_2-C_5-C_5N-C_6-C_6
C_{20} Ochrobirine

C_4N-C_4N-C_5-C_5O-C_6-C_6
C_{20} Gelsemine

5,5,6/5,5,6

C_4N-C_4N-C_6/C_4N-C_4N-C_6
C_{23} Calycanthidine

5,5,6,6,6,6

C_3NO-C_4N-C_6-C_6-C_6-C_6
C_{22} Isoatisine

C_3O_2-C_3O_2-C_5N-C_6-C_6
C_{20} Chelidonine

C_4N-C_4N-C_5N-C_6-C_6-C_6
C_{19} Aspidofractinine

C_4N-C_5-C_5N-C_5N-C_5N-C_6
C_{20} Ajmalan

C_4N-C_5-C_5N-C_6-C_6-C_6
C_{21} Himbosan
C_{27} Solanidane

C_4O-C_5-C_5N-C_6-C_6-C_6
C_{27} Jervanine
Spirosolane

C_4O-C_5-C_5O-C_6-C_6-C_6
C_{27} Spirostan

C_5-C_5-C_5N-C_6-C_6-C_6
C_{18} Aconitane

5,6,6/6,6,6

C_4N-C_5N-C_6/C_5N-C_5N-C_6
C_{27} Tubulosan

5,6,6,6,6,6

C_3NO-C_5N-C_6-C_6-C_6-C_6
C_{22} Atisine

C_5-C_5N-C_5N-C_6-C_6-C_6
C_{27} Cevane

5,6,6,6,7,7

C_5-C_5N-C_5O-C_6-C_6O-C_7
C_{18} Heteratisane

6,6,6,6,6,6

C_4N_2-C_5N-C_5N-C_5N-C_5N-C_6
C_{20} Panamine
C_{21} Homoormosanine

C_4O_2-C_5O-C_6-C_6-C_6-C_6
C_{30} Officinalic acid

C_5N-C_5N-C_5N-C_5N-C_6-C_6
C_{22} Calycanthine

Systems with 7 rings

5,5,5,6,6,6,6

C_4N-C_4N-C_5-C_5N-C_6-C_6-C_6
C_{20} Kopsan

C_4N-C_4N-C_5-C_6-C_6-C_6-C_6
C_{20} Hetisan

5,5,6/5,6,6,6

C_3NO-C_4N-C_4N_2/C_4N-C_5N-C_6-C_6
C_{25} Ergotaman

5,5,6,6,6,6,6

C_5-C_5-C_6-C_6-C_6-C_6-C_6
C_{26} Rubicene

5,5,6,6,6,6,7

C_4N-C_4N-C_5N-C_5N-C_6-C_6-C_6O
C_{21} Strychnidine

6,6,6,6,6,6,6

C_4N_2-C_6-C_6-C_6-C_6-C_6-C_6
C_{28} Anthrazine
Phenanthrazine

C_6-C_6-C_6-C_6-C_6-C_6-C_6
C_{24} Coronene
C_{30} Trinaphthylene

6,6,6,6,6,6,16

C_5N-C_5N-C_6-C_6-C_6-C_6-$C_{15}O$
C_{32} Rodiasan

6,6,6,6,6,6,18

C_5N-C_5N-C_6-C_6-C_6-C_6-$C_{16}O_2$
C_{32} Berbaman
Cycleanan
Oxyacanthan
Thalidasan
Tubocuraran

6,6,6,6,6,6,19

C_5N-C_5N-C_6-C_6-C_6-C_6-$C_{17}O_2$
C_{32} Thalicberan
Thalman

Systems with 8 rings

5,5,6,6,6,6,6,6

C_5-C_5-C_6-C_6-C_6-C_6-C_6-C_6
C_{30} Rugulosin

6,6,6,6,6,6,6,6

C_6-C_6-C_6-C_6-C_6-C_6-C_6-C_6
C_{30} Pyranthrene

6,6,6,6,6,6,7,18

C_5N-C_5N-C_6-C_6-C_6-C_6-C_5O_2-$C_{16}O_2$
C_{38} Insularine

Systems with 9 rings

5,5,5,5,6,6,6,6,16

C_4N-C_4N-C_4N-C_4N-C_6-C_6-C_6-C_6-C_8N_8
C_{32} Phthalocyanine

C_4N-C_4N-C_4N-C_4N-C_6-C_6-C_6-C_6-$C_{12}N_4$
C_{36} Tetrabenzoporphyrin

6,6,6,6,6,6,6,6,6

C_6-C_6-C_6-C_6-C_6-C_6-C_6-C_6-C_6
C_{34} Isoviolanthrene
Violanthrene

Systems with 10 rings

5,5,5,6,6,6,6,6,6,7

C_4N-C_4O-C_5-C_5N-C_5N-C_6-C_6-C_6-C_6-C_6O
C_{36} Cancentrine

6,6,6,6,6,6,6,6,6,6

C_6-C_6-C_6-C_6-C_6-C_6-C_6-C_6-C_6-C_6
C_{32} Ovalene

Systems with 11 rings

5,5,5,5,6,6,6,6,6,6,8

C_4N-C_4N-C_4N-C_4N-C_5N-C_5N-C_6-C_6-C_6-C_6-C_6N_2
C_{40} Toxiferin I

Systems with 12 rings

5,5,5,5,5,5,6,6,6,6,6,6

C_4N-C_4N-C_4N-C_4N-C_4N-C_4N-C_5N-C_5N-C_6-C_6
C_{40} *C*-Calebassine

Systems with 13 rings

5,5,5,5,5,5,6,6,6,6,6,6,6

C_4N-C_4N-C_4N-C_4N-C_4O-C_4O-C_5N-C_5N-C_5N-C_6-C_6-C_6-C_6
C_{43} Vobtusine

5,5,6,6,6,6,6,6,6,6,6,6,6

C_5-C_5-C_5N-C_5N-C_5O-C_6-C_6-C_6-C_6-C_6-C_6-C_6-C_6
C_{41} Staphimine

Systems with 14 rings

5,5,5,5,5,5,6,6,6,6,6,6,7,7

C_4N-C_4N-C_4N-C_4N-C_4N-C_4N-C_5N-C_5N-C_6-C_6-C_6O-C_6O
C_{38} Caracurine II

Formula-Index

C_1

CH_2N_2
Carbodiimide
CH_2O
Formaldehyde
CH_2O_2
Formic acid
CH_2O_3
Carbonic acid
CH_3NO_2
Carbamic acid
CH_4
Methane
CH_4N_2O
Isourea
Urea
$CH_4N_2O_2$
Carbazic acid
CH_5N_3
Guanidine
CH_5N_3O
Semicarbazide

C_2

C_2H_2
Acetylene
$C_2H_2N_2O$
Furazan
C_2H_2O
Ketene
$C_2H_2O_2$
Glyoxal
$C_2H_2O_3$
Glyoxylic acid
$C_2H_2O_4$
Oxalic acid
$C_2H_3NO_3$
Oxamic acid
$C_2H_4N_2O_3$
Allophanic acid
$C_2H_4N_2S_3$
Thiuram sulfide
$C_2H_4N_2S_4$
Thiuram disulfide
C_2H_4O
Acetaldehyde
$C_2H_4O_2$
Acetic acid
$C_2H_4O_3$
Glycolic acid
$C_2H_5NO_2$
Glycine
$C_2H_5N_3OS$
Thiobiuret
$C_2H_5N_3O_2$
Biuret
$C_2H_5N_3S_2$
Dithiobiuret
C_2H_6
Ethane
$C_2H_6O_2$
Ethylene glycol
$C_2H_7NO_3S$
Taurine

C_3

$C_3H_2O_2$
Propiolic acid

$C_3H_2O_5$
Mesoxalic acid
C_3H_3NO
Isoxazole
$C_3H_3NOS_2$
Rhodanine
C_3H_3NS
Isothiazole
C_3H_4
Allene
$C_3H_4N_2$
Imidazole
Pyrazole
$C_3H_4N_2O_2$
Hydantoin
$C_3H_4O_2$
Acrylic acid
$C_3H_4O_3$
Pyruvic acid
$C_3H_4O_4$
Malonic acid
$C_3H_4O_5$
Tartronic acid
$C_3H_6N_2$
Imidazoline
Pyrazoline
$C_3H_6N_2O_3$
Hydantoic acid
$C_3H_6N_4O_3$
Triuret
C_3H_6O
Acetone
$C_3H_6O_2$
Propionic acid
$C_3H_6O_3$
Lactic acid
$C_3H_6O_4$
Glyceric acid
C_3H_7NO
Isoxazolidine
$C_3H_7NO_2$
Alanine
β-Alanine
Sarcosine
$C_3H_7NO_2S$
Cysteine
$C_3H_7NO_3$
Serine
C_3H_8
Propane
$C_3H_8N_2$
Imidazolidine
Pyrazolidine
$C_3H_8O_3$
Glycerol

C_4

$C_4H_2N_2O_4$
Alloxan
C_4H_4
Tetrahedrane
$C_4H_4N_2$
Pyrazine
Pyridazine
Pyrimidine
$C_4H_4N_2O_2$
Uracil
$C_4H_4N_2O_3$
Barbituric acid
C_4H_4O
Furan
$C_4H_4O_4$
Fumaric acid
Maleic acid
$C_4H_4O_5$
Oxalacetic acid
C_4H_4S
Thiophene
C_4H_5N
Pyrrole
$C_4H_5N_3O$
Cytosine
$C_4H_6O_2$
Crotonic acid
Methacrylic acid
Succinaldehyde
$C_4H_6O_3$
Acetoacetic acid
$C_4H_6O_4$
Succinic acid

$C_4H_6O_5$
Malic acid
$C_4H_6O_6$
Tartaric acid
C_4H_7N
Pyrroline
$C_4H_7NO_4$
Aspartic acid
$C_4H_8N_2O_3$
Asparagine
C_4H_8O
Butyraldehyde
$C_4H_8O_2$
Butyric acid
$C_4H_8O_4$
Erythrose
Threose
C_4H_9N
Pyrrolidine
C_4H_9NO
Morpholine
$C_4H_9NO_2S$
Homocysteine
$C_4H_9NO_3$
Allothreonine
Homoserine
Threonine
C_4H_{10}
Butane
Isobutane
$C_4H_{10}N_2$
Piperazine

C_5

$C_5H_4N_2O_4$
Orotic acid
$C_5H_4N_4$
Purine
$C_5H_4N_4O$
Hypoxanthine
$C_5H_4N_4O_2$
Xanthine
C_5H_5N
Pyridine
$C_5H_5N_5$
Adenine
$C_5H_5N_5O$
Guanine
$C_5H_6N_2O_2$
Thymine
C_5H_6O
Pyran
C_5H_8
Isoprene
$C_5H_8O_4$
Glutaric acid
$C_5H_9NO_2$
Proline
$C_5H_9NO_4$
Glutamic acid
$C_5H_{10}N_2O_3$
Glutamine
$C_5H_{10}O_2$
Valeric acid
$C_5H_{10}O_5$
Apiose
Arabinofuranose
Arabinopyranose
Arabinose
Lyxofuranose
Lyxopyranose
Lyxose
Ribofuranose
Ribopyranose
Ribose
Xylofuranose
Xylopyranose
Xylose
$C_5H_{11}N$
Piperidine
$C_5H_{11}NO_2$
Betaine
Isovaline
Norvaline
Valine
$C_5H_{11}NO_2S$
Methionine
C_5H_{12}
Isopentane
Neopentane

$C_5H_{12}N_2O_2$
Ornithine
$C_5H_{12}O_4$
Pentaerythritol
$[C_5H_{14}NO]^+$
Choline

C_6

$C_6H_3N_3O_8$
Styphnic acid
$C_6H_4N_3O_7$
Picric acid
$C_6H_4N_4$
Pteridine
$C_6H_5NO_2$
Isonicotic acid
Nicotinic acid
C_6H_6
Benzene
Fulvene
C_6H_6O
Phenol
$C_6H_6O_2$
Hydroquinone
Pyrocatechol
Resorcinol
$C_6H_6O_3$
Phloroglucinol
Pyrogallol
C_6H_7N
Aniline
$C_6H_7NO_3S$
Sulfanilic acid
$C_6H_8O_6$
Ascorbic acid
$C_6H_9N_3O_2$
Histidine
$C_6H_{10}O_4$
Adipic acid
$C_6H_{11}NO$
ε-Caprolactam
$C_6H_{12}N_2O_4S$
Lanthionine
$C_6H_{12}N_2O_4S_2$
Cystine
$C_6H_{12}O_5$
Quercitol
$C_6H_{12}O_6$
Allofuranose
Allopyranose
Allose
Altrofuranose
Altropyranose
Altrose
Fructofuranose
Fructopyranose
Fructose
Galactofuranose
Galactopyranose
Galactose
Glucofuranose
Glucopyranose
Glucose
Gulofuranose
Gulopyranose
Gulose
Idofuranose
Idopyranose
Idose
Inositol
Mannofuranose
Mannopyranose
Mannose
Psicose
Psicofuranose
Psicopyranose
Sorbofuranose
Sorbopyranose
Sorbose
Tagatofuranose
Tagatopyranose
Tagatose
Talofuranose
Talopyranose
Talose
$C_6H_{13}NO_2$
Alloisoleucine
Isoleucine
Leucine
Norleucine
C_6H_{14}
Isohexane

$C_6H_{14}N_2O_2$
Lysine
$C_6H_{14}N_2O_4$
Streptamine
$C_6H_{14}N_4O_2$
Arginine

C_7

$C_7H_6N_2$
Indazole
$C_7H_6O_2$
Benzoic acid
$C_7H_6O_3$
Salicylic acid
$C_7H_6O_4$
Protocatechuic acid
$C_7H_6O_5$
Gallic acid
$C_7H_7NO_2$
Anthranilic acid
C_7H_8
Toluene
C_7H_8O
Anisole
Benzyl alcohol
Cresol
$C_7H_8O_2$
Guaiacol
C_7H_9NO
Anisidine
C_7H_{12}
Norbornane
Norcarane
Norpinane
$C_7H_{13}N$
Quinuclidine
$C_7H_{14}N_2O_4S$
Cystathionine
$C_7H_{14}O_7$
Sedoheptulose

C_8

C_8H_6
Pentalene
$C_8H_6N_2$
Cinnoline
Naphthyridine
Phthalazine
Quinazoline
Quinoxaline
C_8H_6O
Isobenzofuran
$C_8H_6O_2$
Phthalide
$C_8H_6O_3$
Piperonal
$C_8H_6O_4$
Isophthalic acid
Phthalic acid
Piperonylic acid
Terephthalic acid
C_8H_7N
Indole
Indolizine
Isoindole
C_8H_8
Cubane
Styrene
C_8H_8O
Phthalan
$C_8H_8O_3$
Anisic acid
Vanillin
$C_8H_8O_4$
Vanillic acid
C_8H_9N
Indoline
Isoindoline
$C_8H_9NO_3$
Pyridoxal
C_8H_{10}
Xylene
$C_8H_{10}O$
Phenethyl alcohol
Phenetole
Xylenol
$C_8H_{10}O_2$
Veratrole
$C_8H_{11}N$
Xylidine

$C_8H_{11}NO$
Phenetidine
$C_8H_{11}NO_3$
Pyridoxine
$C_8H_{11}NO_3S$
Penicillanic acid
$C_8H_{12}N_2O_2$
Pyridoxamine
$C_8H_{15}N$
Tropane

C_9

$C_9H_6O_2$
Coumarin
Isocoumarin
C_9H_7N
Isoquinoline
Quinoline
C_9H_8
Indene
C_9H_8O
Chromene
Isochromene
$C_9H_8O_2$
Cinnamic acid
C_9H_9N
Quinolizine
$C_9H_9NO_3$
Hippuric acid
C_9H_{10}
Indan
$C_9H_{10}O$
Chroman
Isochroman
$C_9H_{10}O_3$
Tropic acid
$C_9H_{10}O_4$
Veratric acid
$C_9H_{11}NO_2$
Phenylalanine
$C_9H_{11}NO_3$
Tyrosine
C_9H_{12}
Cumene
Mesitylene
$C_9H_{12}N_2O_6$
Uridine
$C_9H_{13}N_3O_5$
Cytidine
$C_9H_{17}NO_7$
Muramic acid
$C_9H_{17}NO_8$
Neuramic acid

C_{10}

$C_{10}H_8$
Azulene
Naphthalene
$C_{10}H_8O$
Naphthol
$C_{10}H_9NO_3S$
Naphthionic acid
$C_{10}H_{10}$
Bullvalene
$C_{10}H_{10}O_2$
Isosafrole
Safrole
$C_{10}H_{12}N_2O_8$
Orotidine
$C_{10}H_{12}N_4O_5$
Inosine
$C_{10}H_{12}N_4O_6$
Xanthosine
$C_{10}H_{12}O$
Anethole
$C_{10}H_{12}O_2$
Eugenol
$C_{10}H_{13}N_5O_4$
Adenosine
$C_{10}H_{13}N_5O_5$
Guanosine
$C_{10}H_{14}$
Cymene
$C_{10}H_{14}N_2$
Anabasine
Nicotine
$C_{10}H_{14}N_2O_5$
Thymidine
$C_{10}H_{14}O$
Carvacrol
Thymol

$C_{10}H_{16}$
Adamantane
Camphene
$C_{10}H_{16}O$
Citral
$C_{10}H_{16}O_4$
Camphoric acid
$C_{10}H_{18}$
Bornane
Carane
Pinane
Thujane
$C_{10}H_{18}O$
Geraniol
$C_{10}H_{20}$
Iridane
Menthane
$C_{10}H_{20}O$
Menthol

C_{11}

$C_{11}H_7N_3$
Anthyridine
$C_{11}H_8N_2$
β-Carboline
Perimidine
$C_{11}H_8O_2$
Naphthoic acid
$C_{11}H_{12}N_2O_2$
Tryptophane
$C_{11}H_{14}N_2O$
Cytisine

C_{12}

$C_{12}H_8$
Acenaphthylene
Biphenylene
as-Indacene
s-Indacene
$C_{12}H_8As_2$
Arsanthrene
$C_{12}H_8AsN$
Phenarsazine
$C_{12}H_8N_2$
Phenanthroline
Phenazine
$C_{12}H_8OS$
Phenoxathiin
$C_{12}H_8S_2$
Thianthrene
$C_{12}H_9N$
Carbazole
$C_{12}H_9NO$
Phenoxazine
$C_{12}H_9NS$
Phenothiazine
$C_{12}H_{10}$
Acenaphthene
Biphenyl
Heptalene
$C_{12}H_{10}O_4$
Piperonic acid
$C_{12}H_{12}N_2$
Benzidine

C_{13}

$C_{13}H_9As$
Acridarsine
Arsanthridine
$C_{13}H_9N$
Acridine
Phenanthridine
$C_{13}H_{10}$
Fluorene
Phenalene
$C_{13}H_{10}O$
Xanthene
$C_{13}H_{17}NO$
Securinan

C_{14}

$C_{14}H_{10}$
Anthracene
Phenanthrene

$C_{14}H_{10}O_2$
Benzil
$C_{14}H_{11}N_3$
Tripyrrin
$C_{14}H_{12}$
Stilbene
$C_{14}H_{12}N_6O_3$
Pteroic acid
$C_{14}H_{12}O$
Deoxybenzoin
$C_{14}H_{12}O_3$
Benzilic acid
$C_{14}H_{16}N_2$
Ergoline

C_{15}

$C_{15}H_9N$
Thebenidine
$C_{15}H_{10}$
Trindene
$C_{15}H_{10}N_2$
Quindoline
Quinindoline
$C_{15}H_{12}O$
Chalcone
$C_{15}H_{15}NO_4$
Thyronine
$C_{15}H_{19}N$
Galanthan
$C_{15}H_{19}NO$
Galanthaman
Lycoranan
$C_{15}H_{24}$
Gibbane
$C_{15}H_{25}N$
Lycopodane
$C_{15}H_{26}$
Caryolane
Cedrane
Clovane
Longifolane
$C_{15}H_{26}N_2$
Matridine
Sparteine
$C_{15}H_{26}O$
Apotrichothecane
Trichothecane
$C_{15}H_{28}$
Ambrosane
Bakkane
Cadinane
Drimane
Eremophilane
Eudesmane
Guaiane
$C_{15}H_{30}$
Germacrane

C_{16}

$C_{16}H_{10}$
Aceanthrylene
Acephenanthrylene
Fluoranthene
Pyrene
$C_{16}H_{12}$
Aceanthrene
Acephenanthrene
$C_{16}H_{15}N$
Aporphane
$C_{16}H_{17}NO_4$
Pancracine
$C_{16}H_{19}NO_2$
Crinane
Eleocarpine
$C_{16}H_{21}N$
Erythrinan
Hasubanan
Morphinan
$C_{16}H_{21}NO_3$
Annotinine
$C_{16}H_{22}N_2$
Lycodine
$C_{16}H_{23}NO$
Fawcettidine
$C_{16}H_{25}NO$
Dendrobane

C_{17}

$C_{17}H_{17}N$
Berbine

$C_{17}H_{17}NO$
Rheadan
$C_{17}H_{20}N_4O_6$
Riboflavin
$C_{17}H_{22}N_2$
Dasycarpidan
$C_{17}H_{27}NO_2$
Crotalanan
$C_{17}H_{28}$
Gonane
$C_{17}H_{30}$
Podocarpane

C_{18}

$C_{18}H_{12}$
Chrysene
Naphthacene
Pleiadene
Triphenylene
$C_{18}H_{12}N_2$
Phthaloperine
$C_{18}H_{21}NO_5$
Tazettine
$C_{18}H_{22}N_2O$
Formosanan
$C_{18}H_{24}N_2$
Condyfolan
$C_{18}H_{27}N$
Aconitane
$C_{18}H_{27}NO$
Heteratisane
$C_{18}H_{29}NO_2$
Senecionan
$C_{18}H_{30}$
Estrane
$C_{18}H_{34}O_2$
Elaidic acid
Oleic acid
$C_{18}H_{37}NO_2$
Sphingosine
$C_{18}H_{39}NO_2$
Sphinganine

C_{19}

$C_{19}H_{12}N_2$
Acrindoline
$C_{19}H_{14}N_4$
Bilin
Corrole
$C_{19}H_{20}N_2O$
Meloscine
$C_{19}H_{22}N_2$
Akuammilan
Eburnamenine
$C_{19}H_{22}N_2$
Sarpagane
$C_{19}H_{22}N_4$
Corrin
$C_{19}H_{22}O_2$
Cinchonan
$C_{19}H_{24}ClNO_6$
Acutumine
$C_{19}H_{24}N_2$
Aspidofractinine
Ibogamine
$C_{19}H_{24}N_2O_3$
Gelsedine
$C_{19}H_{26}N_2$
Aspidospermidine
Corynan
Curan
$C_{19}H_{28}N_2$
Corynoxan
Yohimban
$C_{19}H_{31}N$
Atidane
Veatchane
$C_{19}H_{32}$
Androstane

C_{20}

$C_{20}H_{12}$
Perylene
$C_{20}H_{14}$
Cholanthrene
$C_{20}H_{14}N_4$
Porphyrin

$C_{20}H_{16}N_4$
Chlorin
$C_{20}H_{18}N_4$
Bacteriochlorin
$C_{20}H_{19}NO_5$
Chelidonine
$C_{20}H_{19}NO_6$
Ochrobirine
$C_{20}H_{20}N_4$
Porphyrinogen
$C_{20}H_{22}N_2O_2$
Gelsemine
$C_{20}H_{24}N_2$
Kopsan
$C_{20}H_{26}N_2$
Ajmalan
Vobasan
$C_{20}H_{27}N$
Hetisan
$C_{20}H_{28}O$
Retinal
$C_{20}H_{28}O_2$
Retinoic acid
$C_{20}H_{30}$
Deoxyretinol
$C_{20}H_{30}O$
Retinol
$C_{20}H_{32}O_7$
Leuconolide
$C_{20}H_{33}N_3$
Panamine
$C_{20}H_{34}$
Atisane
Grayanotoxane
Kaurane
Phyllocladane
Tigliane
$C_{20}H_{34}O$
Picrasane
$C_{20}H_{34}O_7$
Oleandolide
$C_{20}H_{35}N_3$
Ormosanine
$C_{20}H_{36}$
Clavularane
Abietane
Pimarane
Rosane
Taxane
$C_{20}H_{37}N$
[11]Cytochalasan
$C_{20}H_{38}$
Clerodane
Labdane
$C_{20}H_{40}$
Lobane
Prostane
$C_{20}H_{40}O$
Phytol
Thromboxane

C_{21}

$C_{21}H_{24}N_2O$
Strychnidine
$C_{21}H_{24}N_2O_4$
Talbotine
$C_{21}H_{26}N_2O$
Alstophyllan
$C_{21}H_{31}N$
Himbosan
$C_{21}H_{35}N_3$
Homoormosanine
$C_{21}H_{36}$
Diginane
Pregnane

C_{22}

$C_{22}H_{14}$
Picene
$C_{22}H_{18}N_4$
Phorbine
$C_{22}H_{26}N_4$
Calycanthine
$C_{22}H_{33}NO_2$
Atisine
Isoatisine
$C_{22}H_{33}NO_8$
Parsonsine

$C_{22}H_{37}N$
Conane
Daphnane
$C_{22}H_{41}N$
[13]Cytochalasan

C_{23}

$C_{23}H_{28}N_4$
Calycanthidine
$C_{23}H_{36}O_2$
Cardanolide

C_{24}

$C_{24}H_{12}$
Coronene
$C_{24}H_{17}N_5$
Sapphyrin
$C_{24}H_{27}NO$
Lythran
$C_{24}H_{34}O_8$
Caryoptinol
$C_{24}H_{38}O_2$
Bufanolide
$C_{24}H_{42}$
Cholane

C_{25}

$C_{25}H_{31}N_3O_4$
Lunaridine
Lunarine
$C_{25}H_{32}N_2$
Emetan
$C_{25}H_{33}N_5O$
Ergotaman
$C_{25}H_{46}$
Ophiobolane

C_{26}

$C_{26}H_{14}$
Rubicene
$C_{26}H_{32}O_8$
Obacunoic acid
$C_{26}H_{34}O_{10}$
Limonoic acid
$C_{26}H_{44}O_2$
Gedunan
Tocol

C_{27}

$C_{27}H_{33}N_3$
Tubulosan
$C_{27}H_{43}N$
Veratraman
$C_{27}H_{44}O$
Calciol
Isocalciol
Tacalciol
$C_{27}H_{44}O_2$
Spirostan
$C_{27}H_{45}N$
Cevane
Solanidane
$C_{27}H_{45}NO$
Jervanine
Spirosolane
$C_{27}H_{46}$
Fesane
$C_{27}H_{46}O$
Furostan
$C_{27}H_{47}N$
Veratran
$C_{27}H_{48}$
Cholestane
Jervane

C_{28}

$C_{28}H_{16}N_2$
Anthrazine
Phenanthrazine
$C_{28}H_{44}O$
Ercalciol
$C_{28}H_{50}$
Ergostane
Lumistane

$C_{28}H_{50}N_2O_4$
Carpaine

C_{29}

$C_{29}H_{52}$
Stigmastane

C_{30}

$C_{30}H_{16}$
Pyranthrene
$C_{30}H_{18}$
Trinaphthylene
$C_{30}H_{22}O_{10}$
Rugulosin
$C_{30}H_{30}N_4O_4$
Deuteroporphyrin
$C_{30}H_{44}O_6$
Officinalic acid
$C_{30}H_{50}$
Squalene
$C_{30}H_{52}$
Friedelane
Gammacerane
Gorgostane
Lupane
Oleanane
Taraxastane
Taraxerane
Ursane
$C_{30}H_{54}$
Cucurbitane
Dammarane
Euphane
Lanostane
Onocerane
Tirucallane
$C_{30}H_{56}$
Ambrane

C_{31}

$C_{31}H_{34}N_4O_2$
Pyrroporphyrin
$C_{31}H_{36}N_4O_2$
Pyrrochlorin
$C_{31}H_{56}$
Eburicane
Euphorbane

C_{32}

$C_{32}H_{14}$
Ovalene
$C_{32}H_{18}N_8$
Phthalocyanine
$C_{32}H_{30}N_2O$
Rodiasan
$C_{32}H_{30}N_2O_2$
Berbaman
Cycleanan
Oxyacanthan
Thalicberan
Thalidasan
Thalman
Tubocuraran
$C_{32}H_{34}N_4O_4$
Rhodoporphyrin
$C_{32}H_{36}N_4O_2$
Phylloporphyrin
$C_{32}H_{36}N_4O_4$
Rhodochlorin
$C_{32}H_{38}N_4$
Etioporphyrin I
$C_{32}H_{38}N_4O_2$
Phyllochlorin

C_{33}

$C_{33}H_{34}N_4O_3$
Phytoporphyrin
$C_{33}H_{34}N_4O_6$
Biliverdin
$C_{33}H_{36}N_4O_3$
Phytochlorin
$C_{33}H_{36}N_4O_6$
Bilirubin
$C_{33}H_{38}N_4O_6$
Mesobiliverdin

$C_{33}H_{40}N_4O_6$
Mesobilirubin
$C_{33}H_{42}N_4O_6$
Urobilin
$C_{33}H_{44}N_4O_6$
Urobilinogen
$C_{33}H_{46}N_4O_6$
Stercobilin

C_{34}

$C_{34}H_{20}$
Isoviolanthrene
Violanthrene
$C_{34}H_{34}N_4O_4$
Protoporphyrin
$C_{34}H_{38}N_4O_4$
Mesoporphyrin
$C_{34}H_{38}N_4O_6$
Hematoporphyrin
$C_{34}H_{46}ClN_3O_{10}$
Maytansine

C_{35}

$C_{35}H_{34}N_4O_6$
Pheophorbide b
$C_{35}H_{36}N_4O_5$
Pheophorbide a

C_{36}

$C_{36}H_{22}N_4$
Tetrabenzoporphyrin
$C_{36}H_{34}N_2O_7$
Cancentrine
$C_{36}II_{38}N_4O_8$
Coproporphyrin I
$C_{36}H_{43}NO_{17}$
Evonimine
Evonine

C_{38}

$C_{38}H_{38}N_4O_2$
Caracurine II
$C_{38}H_{40}N_2O_6$
Insularine

C_{39}

$C_{39}H_{49}N$
O_{14}
Streptovaricinoic acid

C_{40}

$C_{40}H_{38}N_4O_{16}$
Uroporphyrin I
$[C_{40}H_{46}N_4O_2]^{2+}$
Toxiferin I
$C_{40}H_{48}$
φ,φ-Carotene
φ,χ-Carotene
χ,χ-Carotene
$[C_{40}H_{48}N_4O_2]^{2+}$
C-Calebassine
$C_{40}H_{52}$
β,φ-Carotene
β,χ-Carotene
φ,ψ-Carotene
$C_{40}H_{54}$
κ,φ-Carotene
κ,χ-Carotene
$C_{40}H_{56}$
β,β-Carotene
β,ε-Carotene
β,ψ-Carotene
ε,ε-Carotene
ε,ψ-Carotene
ψ,ψ-Carotene
$C_{40}H_{58}$
β,κ-Carotene
$C_{40}H_{60}$
κ,κ-Carotene
$C_{40}H_{64}O_{12}$
Nonactin

C_{41}

$C_{41}H_{54}N_2O$
Staphimine

C_{43}

$C_{43}H_{50}N_4O_6$
Vobtusine

C_{49}

$C_{49}H_{58}N_4O_6$
Cytoporphyrin

Reg.No.-Index

50-00-0 Formaldehyde
50-14-6 Ercalciol
50-21-5 Lactic acid
50-69-1 D-Ribose
50-71-5 Alloxan
50-81-7 L-Ascorbic acid
50-89-5 Thymidine
50-99-7 D-Glucose

52-90-4 L-Cysteine

54-11-5 Nicotine
54-12-6 DL-Tryptophane

55-22-1 Isonicotic acid

56-40-6 Glycine
56-41-7 L-Alanine
56-45-1 L-Serine
56-81-5 Glycerol
56-84-8 L-Aspartic acid
56-85-9 L-Glutamine
56-86-0 L-Glutamic acid
56-87-1 L-Lysine
56-88-2 L,L-Cystathionine
56-89-3 L-Cystine

57-13-6 Urea
57-48-7 D-Fructose
57-56-7 Semicarbazide

58-61-7 Adenosine
58-63-9 Inosine
58-86-6 D-Xylose
58-96-8 Uridine

59-23-4 D-Galactose
59-51-8 DL-Methionine
59-67-6 Nicotinic acid

60-12-8 Phenethyl alcohol
60-18-4 L-Tyrosine

61-90-5 L-Leucine

62-49-7 Choline
62-53-3 Aniline

63-68-3 L-Methionine
63-91-2 L-Phenylalanine

64-18-6 Formic acid
64-19-7 Acetic acid

65-23-6 Pyridoxine
65-42-9 Lyxose
65-46-3 Cytidine
65-71-4 Thymine
65-85-0 Benzoic acid
65-86-1 Orotic acid

66-22-8 Uracil
66-71-7 [1,10]Phenanthroline
66-72-8 Pyridoxal

67-52-7 Barbituric acid
67-64-1 Acetone
67-97-0 Calciol

68-26-8 Retinol
68-94-0 Hypoxanthine

69-72-7 Salicylic acid
69-89-6 Xanthine

70-26-8 L-Ornithine
70-47-3 L-Asparagine
70-54-2 DL-Lysine

71-00-1 L-Histidine
71-30-7 Cytosine
71-43-2 Benzene

72-18-4 L-Valine
72-19-5 L_s-Threonine

73-22-3 L-Tryptophane
73-24-5 Adenine
73-32-5 L-Isoleucine
73-40-5 Guanine

74-79-3 L-Arginine
74-82-8 Methane
74-84-0 Ethane
74-86-2 Acetylene
74-98-6 Propane

75-07-0 Acetaldehyde
75-28-5 Isobutane

76-93-7 Benzilic acid

77-95-2 (–)-Quinic acid

78-78-4 Isopentane
78-79-5 Isoprene

79-09-4 Propionic acid
79-10-7 Acrylic acid
79-14-1 Glycolic acid
79-33-4 L-Lactic acid
79-41-4 Methacrylic acid
79-92-5 Camphene

80-68-2 DL-Threonine
80-69-3 Tartronic acid

81-31-2 Violanthrene

82-71-3 Styphnic acid

83-32-9 Acenaphthene
83-88-5 Riboflavin

84-86-6 Naphthionic acid

85-01-8 Phenanthrene
85-87-0 Pyridoxamine

86-55-5 [1]Naphthoic acid
86-73-7 Fluorene
86-74-8 9*H*-Carbazole

87-41-2 Phthalide
87-53-6 Penicillanic acid
87-59-2 [2,3]Xylidine
87-62-7 [2,6]Xylidine
87-66-1 Pyrogallol
87-69-4 L_g-Tartaric acid
87-79-6 L-Sorbose
87-81-0 D-Tagatose
87-89-8 *myo*-Inositol

88-89-1 Picric acid
88-99-3 Phthalic acid

89-78-1 Menthol
89-83-8 Thymol

90-04-0 *o*-Anisidine
90-05-1 Guaiacol
90-15-3 [1]Naphthol
90-39-1 Sparteine

91-16-7 Veratrole
91-18-9 Pteridine
91-19-0 Quinoxaline
91-20-3 Naphthalene
91-22-5 Quinoline
91-64-5 Coumarin

92-24-0 Naphthacene
92-52-4 Biphenyl
92-82-0 Phenazine
92-83-1 9*H*-Xanthene
92-84-2 10*H*-Phenothiazine
92-85-3 Thianthrene
92-87-5 Benzidine

93-07-2 Veratric acid
93-09-4 [2]Naphthoic acid

94-41-7 Chalcone
94-53-1 Piperonylic acid
94-59-7 Safrole
94-70-2 *o*-Phenetidine

95-13-6 1*H*-Indene
95-43-2 D-Threose
95-44-3 L-Threose
95-47-6 *o*-Xylene
95-48-7 *o*-Cresol
95-64-7 [3,4]Xylidine
95-65-8 [3,4]Xylenol
95-68-1 [2,4]Xylidine
95-78-3 [2,5]Xylidine
95-87-4 [2,5]Xylenol

97-53-0 Eugenol
97-67-6 L-Malic acid

98-82-8 Cumene

99-50-3 Protocatechuic acid
99-82-1 *p*-Menthane
99-87-6 *p*-Cymene

100-09-4 *p*-Anisic acid
100-21-0 Terephthalic acid
100-42-5 Styrene
100-51-6 Benzyl alcohol
100-66-3 Anisole
100-76-5 Quinuclidine
101-60-0 Porphyrin
101-66-6 Thyronine
102-94-3 *cis*-Cinnamic acid
103-30-0 *trans*-Stilbene
103-73-1 Phenetole
104-46-1 Anethole
104-94-9 *p*-Anisidine
105-60-2 ε-Caprolactam
105-67-9 [2,4]Xylenol
106-24-1 Geraniol
106-42-3 *p*-Xylene
106-44-5 *p*-Cresol
106-97-8 Butane
107-21-1 Ethylene glycol
107-22-2 Glyoxal
107-35-7 Taurine
107-43-7 Betaine
107-83-5 Isohexane
107-92-6 Butyric acid
107-93-7 *trans*-Crotonic acid
107-95-9 β-Alanine
107-97-1 Sarkosine
108-19-0 Biuret
108-38-3 *m*-Xylene
108-39-4 *m*-Cresol
108-46-3 Resorcinol
108-67-8 Mesitylene
108-68-9 [3,5]Xylenol
108-69-0 [3,5]Xylidine
108-73-6 Phloroglucinol
108-88-3 Toluene
108-95-2 Phenol
109-52-4 Valeric acid
109-97-7 1*H*-Pyrrole

110-00-9 Furan
110-02-1 Thiophene
110-15-6 Succinic acid
110-16-7 Maleic acid
110-17-8 Fumaric acid
110-85-0 Piperazine
110-86-1 Pyridine
110-89-4 Piperidine
110-91-8 Morpholine
110-94-1 Glutaric acid
111-02-4 Squalene
112-79-8 Elaidic acid
112-80-1 Oleic acid
113-00-8 Guanidine
114-04-5 Neuramic acid
114-25-0 Biliverdin
115-77-5 Pentaerythritol
116-31-4 Retinal
118-00-3 Guanosine
118-92-3 Anthranilic acid
119-24-4 Pteroic acid
119-65-3 Isoquinoline
119-98-2 Tocol

120-12-7 Anthracene
120-57-0 Piperonal

120-58-1 Isosafrole
120-72-9 1*H*-Indole
120-73-0 1*H*-Purine
120-80-9 Pyrocatechol
121-33-5 Vanillin
121-34-6 Vanillic acid
121-57-3 Sulfanilic acid
121-91-5 Isophthalic acid
123-31-9 Hydroquinone
123-72-8 Butyraldehyde
123-75-1 Pyrrolidine
123-78-4 Sphingosine
124-04-9 Adipic acid
124-83-4 (1*R*)-*cis*-Camphoric acid
127-17-3 Pyruvic acid
129-00-0 Pyrene

133-37-9 DL-Tartaric acid
134-81-6 Benzil
135-19-3 [2]Naphthol
135-67-1 10*H*-Phenoxazine

140-10-3 *trans*-Cinnamic acid
141-82-2 Malonic acid
141-84-4 Rhodanine
144-62-7 Oxalic acid
144-98-9 DL-Allothreonine
146-80-5 Xanthosine
147-71-7 D_g-Tartaric acid
147-73-9 *meso*-Tartaric acid
147-81-9 Arabinose
147-85-3 L-Proline
149-91-7 Gallic acid

150-30-1 DL-Phenylalanine
150-86-7 Phytol
151-51-9 Carbodiimide
153-94-6 D-Tryptophane
156-43-4 *p*-Phenetidine
157-06-2 D-Arginine
157-39-1 Tetrahedrane

190-26-1 Ovalene
191-07-1 Coronene
191-13-9 Pyranthrene
194-03-6 Thebenidine
196-62-3 Trinaphthylene
197-61-5 Rubicene
198-55-0 Perylene

201-06-9 Acephenanthrylene
202-03-9 Aceanthrylene
203-80-5 Phenalene
204-02-4 Perimidine
206-44-0 Fluoranthene
206-92-8 Pleiadene
208-96-8 Acenaphthylene

210-65-1 *as*-Indacene
213-46-7 Picene
215-14-5 Phenanthrazine
217-59-4 Triphenylene
218-01-9 Chrysene

222-64-0 Anthrazine
228-29-5 3a*H*-Trindene
228-30-8 1*H*-Trindene
229-63-0 Arsanthridine
229-70-9 [3,8]Phenanthroline
229-87-8 Phenanthridine

230-07-9 [4,7]Phenanthroline
230-30-8 [1,8]Phenanthroline
230-45-5 [1,9]Phenanthroline
230-46-6 [1,7]Phenanthroline

243-38-9 5*H*-Quinindoline
243-58-3 11*H*-Quindoline
244-57-5 1*H*-Carbazole
244-63-3 β-Carboline

250-25-9 Pentalene
253-35-0 Isochromene
253-45-2 [2,7]Naphthyridine
253-50-9 [2,6]Naphthyridine
253-52-1 Phthalazine
253-66-7 Cinnoline
253-69-0 [1,7]Naphthyridine
253-72-5 [1,6]Naphthyridine
253-82-7 Quinazoline
254-04-6 Chromene
254-60-4 [1,8]Naphthyridine
254-79-5 [1,5]Naphthyridine
255-58-3 Quinolizine

257-24-9 Heptalene
259-79-0 Biphenylene

260-22-0 Arsanthrene
260-23-1 Acridarsine
260-24-2 Phenarsazine
260-94-6 Acridine
261-15-4 Anthyridine
261-21-2 3*H*-Xanthene
261-76-7 1*H*-Phenoxazine
261-77-8 3*H*-Phenoxazine
261-86-9 1*H*-Phenothiazine
262-20-4 Phenoxathiin
262-76-0 Corrin
267-21-0 *s*-Indacene

270-51-9 3a*H*-Indene
270-52-0 5*H*-Indene
270-53-1 2*H*-Indene
270-55-3 4a*H*-[2]Pyrindine
270-56-4 1*H*-[2]Pyrindine
270-57-5 3*H*-[2]Pyrindine
270-58-6 2*H*-[2]Pyrindine
270-59-7 6*H*-[2]Pyrindine
270-60-0 5*H*-[2]Pyrindine
270-68-8 2*H*-Isoindole
270-69-9 1*H*-Isoindole
270-75-7 Isobenzofuran
270-87-1 4a*H*-[1]Pyrindine
270-88-2 1*H*-[1]Pyrindine
270-89-3 4*H*-[1]Pyrindine
270-90-6 2*H*-[1]Pyrindine
270-91-7 5*H*-[1]Pyrindine
270-92-8 7*H*-[1]Pyrindine
271-23-8 3a*H*-Indole
271-24-9 4*H*-Indole
271-25-0 2*H*-Indole
271-26-1 3*H*-Indole
271-38-5 3a*H*-Indazole
271-39-6 7*H*-Indazole
271-40-9 5*H*-Indazole
271-41-0 4*H*-Indazole
271-42-1 2*H*-Indazole
271-43-2 3*H*-Indazole
271-44-3 1*H*-Indazole
274-40-8 Indolizine
275-51-4 Azulene
277-10-1 Cubane
279-23-2 Norbornane

281-23-2 Adamantane
286-08-8 Norcarane
286-34-0 Norpinane
287-95-6 2*H*-Pyrrole
287-97-8 3*H*-Pyrrole

288-11-9 4*H*-Pyrazole
288-12-0 3*H*-Pyrazole
288-13-1 1*H*-Pyrazole
288-14-2 Isoxazole
288-16-4 Isothiazole
288-32-4 Imidazole
288-37-9 Furazan
289-65-6 4*H*-Pyran
289-66-7 2*H*-Pyran
289-80-5 Pyridazine
289-95-2 Pyrimidine

290-37-9 Pyrazine
298-12-4 Glyoxylic acid

302-72-7 DL-Alanine
302-79-4 Retinoic acid
302-84-1 DL-Serine

312-84-5 D-Serine
314-50-1 Orotidine
319-78-8 D-Isoleucine

327-56-0 D-Norleucine
327-57-1 L-Norleucine
328-38-1 D-Leucine
328-39-2 DL-Leucine
328-42-7 Oxalacetic acid

338-69-2 D-Alanine

344-25-2 D-Proline
348-66-3 D-Ornithine
348-67-4 D-Methionine
349-46-2 D-Cystine

351-50-8 D-Histidine

432-70-2 β,ε-Carotene
438-22-2 (5α)-Androstane
438-23-3 (5β)-Androstane

443-79-8 DL-Isoleucine
448-65-7 Deuteroporphyrin
448-71-5 Etioporphyrin I

451-40-1 Deoxybenzoin
454-28-4 Homocysteine
454-29-5 DL-Homocysteine

461-72-3 Hydantoin
462-60-2 Hydantoic acid
463-49-0 Allene
463-51-4 Ketene
463-77-4 Carbamic acid
463-79-6 Carbonic acid
463-82-1 Neopentane
464-15-3 Bornane
464-60-8 Kopsan
464-93-7 Ursane
464-99-3 Lupane
465-58-7 Isovaline
466-43-3 Atisine
468-10-0 Morphinan
468-83-7 Ambrane
469-84-1 Phyllocladane
469-91-0 Clovane

470-15-5 α-L-Sorbopyranose
470-23-5 β-D-Fructofuranose
470-24-6 β-D-Psicofuranose
471-12-5 Thujane
471-25-0 Propiolic acid
471-31-8 Carbazic acid
471-47-6 Oxamic acid
471-67-0 Oleanane
471-78-3 Podocarpane
472-92-4 ε,ψ-Carotene
472-93-5 β,ψ-Carotene
473-11-0 Eudesmane
473-55-2 Pinane
473-81-4 Glyceric acid
473-90-5 Mesoxalic acid
474-20-4 Lanostane
475-19-4 Longifolane
476-32-4 Chelidonine
478-88-6 Ergoline
479-23-2 Cholanthrene

481-26-5 (5β)-Pregnane
481-87-8 Ibogamine
482-75-7 Cevane
483-49-8 Berbine
483-73-8 Cadinane
484-86-6 Caryolane
485-35-8 Cytisine
488-52-8 Streptamine
488-54-0 *neo*-Inositol
488-55-1 *muco*-Inositol
488-58-4 *epi*-Inositol
488-59-5 *scyllo*-Inositol
488-73-3 Quercitol
489-20-3 Iridane
489-80-5 Guaiane

491-31-6 Isocoumarin
492-61-5 β-D-Glucopyranose
492-62-6 α-D-Glucopyranose
492-66-0 α-L-Glucopyranose
493-05-0 Isochroman
493-08-3 Chroman
493-90-3 Mesoporphyrin
494-52-0 (−)-Anabasine
495-69-2 Hippuric acid
496-11-7 Indan
496-12-8 Isoindoline
496-14-0 Phthalan
496-15-1 Indoline
496-98-0 Lanthionine
497-20-1 Fulvene
498-18-0 Norvaline
498-19-1 Homoserine
499-75-2 Carvacrol

502-65-8 ψ,ψ-Carotene
503-64-0 *cis*-Crotonic acid
503-82-2 Thiuram monosulfide
504-70-1 Pyrazolidine
504-72-3 Isoxazolidine
504-74-5 Imidazolidine
504-90-5 Thiuram disulfide
507-79-9 Tazettine
509-15-9 Gelsemine

510-38-3 Isoatisine
510-70-3 Crinan
511-03-5 Onocerane
512-20-9 α-D-Tagatopyranose
516-06-3 DL-Valine
516-29-0 Euphane
517-30-6 Eburnamenine
519-01-7 Aporphane

523-06-8 Yohimban
524-01-6 φ,φ-Carotene
526-75-0 [2,3]Xylenol
526-83-0 Tartaric acid
527-84-4 *o*-Cymene
529-17-9 Tropane
529-64-6 Tropic acid

531-12-4 Rhodoporphyrin
531-14-6 Coproporphyrin I
533-49-3 L-Erythrose
535-34-2 DL,DL-Cystathionine
535-77-3 *m*-Cymene
536-90-3 *m*-Anisidine

541-50-4 Acetoacetic acid
541-53-7 Dithiobiuret
545-22-2 Dammarane
548-98-1 Cholane
549-07-5 Insularine

550-29-8 φ,χ-Carotene
551-68-8 D-Psicose
551-72-4 L-chiro-Inositol
552-63-6 (±)-Tropic acid
553-12-8 Protoporphyrin
554-59-6 Carane
556-02-5 D-Tyrosine
556-40-6 DL-Tyrosine
556-99-0 Triuret
559-49-9 Annotinine
559-65-9 Gammacerane
559-73-9 Friedelane

560-05-4 (±)-*cis*-Camphoric acid
561-90-0 Labdane
562-08-3 Solanidane

564-86-3 Deoxyretinol
569-24-4 Matridine

574-93-6 Phthalocyanine
576-26-1 [2,6]Xylenol
576-63-6 *cis*-Inositol
579-75-9 *o*-Anisic acid

583-50-6 D-Erythrose
585-21-7 DL-Glutamine
586-38-9 *m*-Anisic acid
588-59-0 Stilbene

595-05-1 Calycanthine
595-32-4 (1*R*)-*trans*-Camphoric acid
595-39-1 DL-Isovaline
595-40-4 L-Isovaline
598-82-3 L-Lactic acid

600-19-1 DL-Glyceric acid
601-58-1 Stigmastane
607-14-7 Uroporphyrin I
608-45-7 α-D-Arabinopyranose
608-46-8 α-D-Lyxopyranose
608-47-9 β-D-Lyxopyranose
609-06-3 L-Xylose
609-36-9 DL-Proline

616-06-8 DL-Norleucine
616-07-9 DL-Ornithine
617-45-8 DL-Aspartic acid
617-48-1 DL-Malic acid
617-65-2 DL-Glutamic acid

621-33-0 *m*-Phenetidine
621-82-9 Cinnamic acid
625-78-5 Allophanic acid

632-20-2 D_s-Threonine
635-65-4 Bilirubin
636-61-3 D-Malic acid
638-37-9 Succinaldehyde
639-97-4 D-Apiose

640-68-6 D-Valine
641-48-5 Aceanthrene

641-85-0 (5α)-Pregnane
643-10-7 *allo*-Inositol
643-12-9 D-*chiro*-Inositol
644-00-8 Pyrroporphyrin

645-10-3 Germacrane
645-49-8 *cis*-Stilbene

668-87-1 [2,8]Phenanthroline

672-15-1 L-Homoserine
673-06-3 D-Phenylalanine

751-29-1 Obacunoic acid
760-78-1 DL-Norvaline
764-22-7 Sphinganine

921-01-7 D-Cysteine
921-60-8 L-Glucose
922-55-4 L-Lanthionine
922-55-4 L-Lanthionine
922-56-5 *meso*-Lanthionine
923-27-3 D-Lysine
923-32-0 DL-Cystine

1005-51-2 Bullvalene
1034-10-2 DL-Thyronine

1114-34-7 D-Lyxose
1114-41-6 Muramic acid

1300-71-6 Xylenol
1300-73-8 Xylidine
1319-77-3 Cresol
1320-04-3 Naphthoic acid
1321-31-9 Phenetidine
1321-67-1 Naphthol
1330-20-7 Xylene
1335-08-6 Anisic acid

1509-34-8 L-Alloisoleucine
1509-35-9 D-Alloisoleucine
1573-40-6 Kaurane
1596-67-4 L-Thyronine

1605-68-1 Taxane
1669-07-4 Corynan
1678-82-6 *trans-p*-Menthane

1758-51-6 Erythrose
1783-96-6 D-Aspartic acid

1856-98-0 Urobilin

1927-25-9 DL-Homoserine
1949-78-6 L-Lyxose
1949-88-8 L-Altrose
1990-29-0 D-Altrose

2013-12-9 D-Norvaline
2054-81-1 Strychnidine
2058-58-4 D-Asparagine
2152-76-3 Idose
2216-51-5 (1*R*)-Menthol
2448-27-3 Panamine
2460-44-8 β-D-Xylopyranose

2595-97-3 D-Allose
2595-98-4 D-Talose
2676-21-3 Allothreonine
2683-78-5 Bacteriochlorin
2683-84-3 Chlorin
2721-43-9 Pyrazoline
2867-15-4 D,L-Cystathionine
2912-09-6 Aspidospermidine
2932-09-4 φ,ψ-Carotene

3019-74-7 Sedoheptulose
3059-97-0 D-Isovaline
3107-04-8 DL-Alloisoleucine
3130-87-8 DL-Asparagine
3183-08-2 DL-Lanthionine
3242-05-5 Eremophilane
3374-22-9 DL-Cysteine
3458-28-4 D-Mannose
3463-92-1 Carpaine

3615-39-2 Sorbose
3615-56-3 D-Sorbose
3617-44-5 Phenylalanine
3646-73-9 α-D-Galactopyranose
3724-65-0 Crotonic acid
3921-76-4 Sarpagane

4180-23-8 *trans*-Anethole
4205-23-6 D-Gulose

4371-52-2 Cysteine
4382-02-9 β,φ-Carotene
4396-11-6 Porphyrinogen
4427-84-3 Cardanolide
4430-29-6 Isoviolanthrene
4466-64-2 Tetrabenzoporphyrin
4732-76-7 Gonane
4744-36-9 Isourea
4998-57-6 DL-Histidine

5001-21-8 Ormosanine
5157-09-5 Norleucine
5285-18-7 Piperonic acid
5328-37-0 L-Arabinose
5392-40-5 Citral
5516-84-7 Caracurine II
5516-85-8 Calycanthidine
5724-81-2 Pyrroline
5876-24-4 Akuammilan
5934-56-5 L-Idose
5949-01-9 Cinchonan
5951-58-6 Drimane
5959-95-5 D-Glutamine
5965-92-4 D-Lanthionine
5978-95-0 D-Idose

6000-40-4 D-Glyceric acid
6020-39-9 *meso*-Cystine
6027-13-0 L-Homocysteine
6027-14-1 D-Homocysteine
6027-21-0 D-Homoserine
6027-89-0 L-Gulose
6035-50-3 DL-Fructose
6038-51-3 Allose
6069-98-3 *cis-p*-Menthane
6173-22-4 Spirostan
6232-48-0 Acephenanthrene
6477-44-7 L-Apiose

6538-00-7 Lumistane
6600-40-4 L-Norvaline
6748-95-4 β-D-Arabinopyranose
6763-34-4 α-D-Xylopyranose
6805-08-9 χ,χ-Carotene
6812-82-4 Rosane
6833-84-7 Nonactin
6871-25-6 Aspidofractinine
6888-23-9 Toxiferin I
6893-26-1 D-Glutamic acid
6898-94-8 Alanine
6898-95-9 Serine
6899-03-2 Aspartic acid
6899-04-3 Glutamine
6899-05-4 Glutamic acid
6899-06-5 Lysine
6899-07-6 Cystathionine
6902-95-0 Gibbane
6912-86-3 Tryptophane
6915-15-7 Malic acid
6917-35-7 Inositol
6929-17-5 (5α)-Cholane

7004-03-7 Valine
7004-09-3 Isoleucine
7004-12-8 Arginine
7005-03-0 Leucine
7005-18-7 Methionine
7005-20-1 Proline
7006-33-9 Ornithine
7006-34-0 Asparagine
7006-35-1 Histidine
7045-51-4 β-D-Galactofuranose
7096-96-0 Gelsedine

7200-25-1 DL-Arginine
7235-40-7 β,β-Carotene
7257-29-6 *C*-Calebassine
7282-78-2 α-D-Gulopyranose
7282-79-3 α-D-Allopyranose
7282-80-6 α-D-Altropyranose
7282-81-7 α-D-Talopyranose
7282-82-8 α-D-Idopyranose
7283-02-5 β-D-Idopyranose
7283-06-9 α-L-Lyxopyranose
7283-07-0 β-L-Lyxopyranose
7283-08-1 β-D-Gulopyranose
7283-09-2 β-D-Allopyranose
7283-10-5 β-D-Altropyranose
7283-11-6 β-D-Talopyranose
7296-15-3 α-D-Mannopyranose
7296-55-1 α-L-Arabinopyranose
7296-56-2 β-L-Arabinopyranose

7296-58-4 α-L-Xylopyranose
7296-59-5 α-D-Ribopyranose
7296-60-8 β-D-Ribopyranose
7296-61-9 α-L-Ribopyranose
7296-62-0 β-L-Ribopyranose
7296-64-2 β-D-Galactopyranose

7322-30-7 β-L-Xylopyranose
7322-31-8 β-D-Mannopyranose
7635-11-2 L-Allose
7660-25-5 β-D-Fructopyranose
7687-39-0 β-D-Lyxofuranose
7776-48-9 L-Fructose

10030-80-5 L-Mannose
10284-69-2 [2,7]Phenanthroline
10323-20-3 D-Arabinose
10326-41-7 D-Lactic acid
10489-79-9 α-D-Fructofuranose
10489-81-3 α-D-Fructopyranose
10504-35-5 D-Ascorbic acid
10550-80-8 Homoormosanine

12772-65-5 α-L-Galactopyranose
12773-29-4 α-L-Altropyranose
12773-31-8 β-L-Altropyranose
12773-32-9 α-L-Talopyranose
12773-33-0 β-L-Idopyranose
12773-34-1 β-L-Mannopyranose

13078-04-1 (±)-Anabasine
13221-22-2 Arabinofuranose
13567-54-9 Cedrane
13939-70-3 Phylloporphyrin

14459-29-1 Hematoporphyrin
14510-67-9 Hasubanan
14684-37-8 Urobilinogen
14795-83-6 α-D-Xylofuranose
14912-31-3 Fawcettidine
14982-53-7 Cholestane

15356-60-2 (1*S*)-Menthol
15356-70-4 (±)-Menthol
15572-79-9 L-Galactose
15664-29-6 Pheophorbide a
15761-67-8 Ribofuranose

16202-15-6 (*S*)-Tropic acid
16354-64-6 L-Psicose
16568-56-2 Mesobilirubin
16580-23-7 *o*-Menthane
16580-24-8 *m*-Menthane

17066-65-8 *cis-m*-Menthane
17066-66-9 *trans-m*-Menthane
17088-50-5 Acutumine
17126-67-9 (*R*)-Tropic acid
17592-07-3 Tacalciol
17598-81-1 Tagatose
17598-82-2 L-Tagatose

18865-70-6 DL-*chiro*-Inositol

19163-87-2 Gulose
19376-21-7 [3,7]Phenanthroline
19376-27-3 [2,9]Phenanthroline
19407-12-6 (4a*R*)-Abietane
19489-02-2 *cis-o*-Menthane
19489-03-3 *trans-o*-Menthane
19772-79-3 Vobtusine

20074-49-1 β-L-Arabinofuranose
20098-65-1 Ophiobolane
20197-42-6 β-D-Tagatopyranose
20235-19-2 DL-Arabinose
20239-99-0 Pheophorbide b
20316-18-1 Lycodine

21416-14-8 Pancracine

22153-41-9 Limonoic acid

23140-52-5 Psicose
23228-74-2 Thiobiuret
23537-16-8 Rugulosin
23567-25-1 L-Talose
23983-77-9 Officinalic acid

24181-64-4 Ochrobirine
24185-51-1 Lunarine
24259-59-4 L-Ribose

24314-51-0 Meloscine
24379-83-7 Atisane
24384-10-9 Apotrichothecane
24479-76-3 Conane
24645-67-8 Cystine
24706-08-9 Trichothecane
24749-18-6 Ambrosane
24749-37-9 Estrane
24755-35-9 Bilin
24830-94-2 D-Allothreonine
24861-47-0 Phorbine
24887-75-0 Androstane
24887-77-2 Bufanolide
24909-91-9 Pregnane

25162-02-1 Cytoporphyrin
25318-39-2 Ergostane
25375-17-1 Taraxastane
25545-03-3 β-D-Arabinofuranose
25545-04-4 α-D-Lyxofuranose
25679-28-1 *cis*-Anethole
25990-60-7 Xylose

26359-43-3 Phytoporphyrin
26444-09-7 Corrole
26524-12-9 Alloisoleucine
26566-61-0 Galactose
26791-77-5 Pyrrochlorin

27198-07-8 1*H*-Xanthene
27711-98-4 Erythrinan

28299-33-4 Imidazoline
28305-26-2 L-Glyceric acid
28954-12-3 L-Allothreonine

29191-52-4 Anisidine
29477-90-5 Cancentrine
29607-01-0 (±)-*trans*-Camphoric acid
29790-16-7 Mesobiliverdin
29825-68-1 DL-Erythrose
29884-64-8 Threose

30077-17-9 Talose
30237-26-4 Fructose
30257-03-5 Pimarane
30412-16-9 β-D-Glucofuranose
30809-15-5 Talbotine
30891-90-8 Eleocarpine

31103-86-3 Mannose
31982-11-3 D,D-Cystathionine

32445-75-3 α-D-Ribofuranose

33458-64-9 Evonine

34159-88-1 Berbaman
34159-89-2 Corynoxan
34159-90-5 Galanthan
34174-79-3 Curan
34217-90-8 Stercobilin
34340-56-2 Lunaridine
34437-70-2 Rheadan
34466-20-1 Ribose
34471-97-1 Spirosolane
34479-48-6 Oxyacanthan
34479-49-7 Securinan
34486-91-4 Senecionan
34486-92-5 Vobasan
34561-13-2 Tubocuraran
34612-39-0 Ajmalan
34612-40-3 Condyfolan
34612-42-5 Ergotaman
34620-86-5 Lythran
34625-83-7 Lycopodane
34709-90-5 Aconitane
34723-40-5 Formosanan
34783-87-4 Furostan

35413-62-8 Streptovaricinoic acid
35732-97-9 Picrasane
35810-56-1 α-L-Mannopyranose
35831-80-2 β-L-Sorbopyranose
35846-35-8 Maytansine

36104-77-5 Tubulosan
36413-57-7 Prostane
36413-60-2 Quinic acid
36441-92-6 α-D-Tagatofuranose
36468-53-8 β-D-Ribofuranose
36468-68-5 α-L-Sorbofuranose
36468-79-8 α-D-Allofuranose

36468-80-1 β-D-Allofuranose
36468-81-2 β-L-Altrofuranose
36468-82-3 α-D-Galactofuranose
36468-84-5 α-D-Glucofuranose
36505-49-4 β,κ-Carotene
36506-59-9 Emetan
36564-41-7 Gorgostane
36573-86-1 Thalicberan
36573-87-2 Thalman
36573-88-3 Thalidasan
36574-15-9 β-L-Idofuranose
36574-19-3 α-D-Gulofuranose
36574-21-7 α-D-Mannofuranose
36676-50-3 Threonine
36712-49-9 κ,κ-Carotene
36972-23-3 α-L-Mannofuranose
36978-43-5 α-L-Talofuranose

37110-85-3 β-D-Xylofuranose
37388-49-1 α-D-Arabinofuranose
37738-80-0 β-L-Mannofuranose

38029-69-5 α-L-Arabinofuranose
38029-84-4 α-D-Psicopyranose
38776-69-1 Fesane
38894-81-4 ε,ε-Carotene
38982-45-5 DL-Threose
38989-96-7 Alstophyllan

39281-65-7 β-L-Glucopyranose
39281-66-8 α-L-Gulopyranose
39281-67-9 β-L-Gulopyranose
39281-68-0 α-L-Idopyranose
39392-61-5 α-L-Allopyranose
39392-62-6 β-L-Allopyranose
39608-68-9 Crotalanan
39608-70-3 Hetisan
39608-76-9 Dendrobane
39608-81-6 Veratraman
39686-16-3 Daphnane
39865-11-7 Taraxerane
39907-73-8 Grayanotoxane

40461-75-4 β-D-Idofuranose
40461-77-6 β-L-Lyxofuranose
40461-79-8 β-D-Altrofuranose
40461-80-1 α-L-Gulofuranose
40461-83-4 β-L-Talofuranose
40461-85-6 β-D-Psicopyranose
40461-86-7 β-D-Tagatofuranose
40461-89-0 α-L-Lyxofuranose
40550-49-0 β-D-Mannofuranose
40774-73-0 Anabasine
40825-89-6 Galactopyranose

41180-88-5 Dasycarpidan
41247-05-6 DL-Xylose
41487-67-0 α-D-Idofuranose
41546-19-4 β-L-Ribofuranose
41546-20-7 α-L-Ribofuranose
41612-84-4 β-L-Fructopyranose
41758-54-7 Evonimine

41846-86-0 α-L-Glucofuranose
41846-88-2 β-L-Galactofuranose
41846-89-3 α-L-Galactofuranose
41846-91-7 α-L-Altrofuranose
41846-93-9 α-D-Altrofuranose
41846-95-1 β-L-Allofuranose
41846-98-4 β-L-Tagatofuranose
41846-99-5 α-L-Tagatofuranose

41847-02-3 β-L-Sorbofuranose
41847-03-4 α-D-Sorbofuranose
41847-06-7 α-D-Psicofuranose
41847-48-7 β-L-Gulofuranose
41847-50-1 β-D-Gulofuranose
41847-53-4 α-L-Psicopyranose
41847-56-7 α-D-Sorbopyranose
41847-63-6 β-D-Talofuranose
41847-65-8 α-L-Idofuranose

41855-38-3 Heteratisane

41904-73-8 Himbosan
41904-80-7 Veatchan
41904-84-1 Atidane

42127-84-4 Acrindoline
42315-76-4 Phthaloperine
42375-21-3 1*H*-Quindoline
42375-22-4 1*H*-Quinindoline
42607-12-5 Isocalciol
42744-42-3 β-D-Sorbofuranose

42789-83-3 β-L-Galactopyranose
42927-70-8 DL-Apiose

51041-84-0 [13]Cytochalasan
51041-85-1 [11]Cytochalasan
51076-04-1 α-D-Talofuranose
51773-71-8 Clerodane
51953-03-8 9*H*-Purine
53683-47-9 Caryoptinol
54808-89-8 β-L-Talopyranosc
55058-43-0 DL-Ribose
55520-40-6 Tyrosine
55781-80-1 β,χ-Carotene
56144-77-5 κ,χ-Carotene
56145-41-6 Phytochlorin
58282-09-0 Tripyrrin
58367-01-4 DL-Glucose
59588-18-0 Staphimine

61828-46-4 Leuconolide
62624-30-0 DL-Ascorbic acid
65441-59-0 Cucurbitane
65732-90-3 DL-Sorbose
66719-58-2 Thromboxane
67707-87-3 Tigliane
68540-16-9 Oleandolide
69307-90-0 Clavularane

71593-03-8 Lobane
72213-98-0 Parsonsine
73952-11-1 DL-Tagatose
76684-08-7 κ,φ-Carotene
78964-23-5 Diginane
79298-93-4 (±)-Lunaridine
79298-94-5 (±)-Lunarine

Springer

K. Hirayama

The Hirn System

Nomenclature of Organic Chemistry
For Man-to-Machine and Man-to-Man Communication
Principles

1984. IX, 149 pages. Hard cover DM 138,–
ISBN 3-540-15031-5
Distribution rights for Japan and Asia: Maruzen Co., Ltd., Tokyo

Contents: Preamble. – General Provisions. – Unmodified Hydrides: Carbanes (Acylic Hydrocarbones). Cyclocarbanes (Acylic Hydrocarbons). Arenes. Cyclarenes (Corannulenes). Fundamental Noncarbon Isohydrides (Hydrides Other Than Hydrocarbons). – Modifiers: Deleo and Seco. Homo and Nor. Cyclo. Dehydro and Hydro. Hetera Modifiers. Substituents. Coarene, Cocycloarenes and Their Analogs. Spiro. Arena Modifiers and Their Analogs. – Radicals. – Compounds, Free Radicals, Ions, and Radical Ions. – Supplemental Examples. – Index.

The nomenclature of organic compounds has a 200 year history of revisions and additions to suit the ever growing number of known compounds. Accordingly, it is difficult to discern a logical consistency. The nomenclature presented in this book is a totally new, simple and logical naming system using only about a hundred basic natural words. It provides not only names for compounds according to present IUPAC rules, but also for those that cannot be named. The new nomenclature is logically consistent through out, and there is a strict correspondence between names and structures. Hence, it is a logical linear-notation system of chemical structures for man-to-machine communication using natural words as well as a simple and plain nomenclature for verbal and visual man-to-man communication.

Springer-Verlag
Berlin Heidelberg
New York Tokyo